ZAMM

ZEITSCHRIFT FÜR ANGEWANDTE MATHEMATIK UND MECHANIK

INGENIEURWISSENSCHAFTLICHE FORSCHUNGSARBEITEN

BAND 56 1976 HEFT 11

AUS DEM INHALT

HAUPTAUFSÄTZE

AKADEMIE-VERLAG · BERLIN

ZAMM Bd. 56 Nr. 11 S. 449—512 Berlin, November 1976

EVP 18,— M 34115

INHALT

Hauptaufsätze

Kleine Mitteilungen

Wir bitten, **Manuskriptsendungen** zweifach (Original und eine Kopie, sprachlich einwandfrei, Formeln mit Maschine oder in Druckschrift geschrieben) an folgende Anschrift zu richten:

Zeitschrift für Angewandte Mathematik und Mechanik z. Hd. Herrn Prof. Dr. Günter Schmidt
Zentralinstitut für Mathematik und Mechanik an der Akademie der Wissenschaften der DDR
DDR-108 Berlin, Mohrenstraße 39.

Zu den Arbeiten, die als Hauptaufsätze bestimmt sind, ist auf gesondertem Blatt eine Zusammenfassung von 5 bis 10 Zeilen in englischer und (möglichst) deutscher und russischer Sprache beizufügen. Ausführliche Hinweise für die Autoren, um deren strikte Berucksichtigung gebeten wird, finden sich im Anschluß an das Inhaltsverzeichnis des Jahrganges **55** (1975).

Die Autoren erhalten von den Hauptaufsatzen 75, von den Kleinen Mitteilungen 25 Sonderdrucke ohne Berechnung, darüber hinaus weitere Sonderdrucke gegen Berechnung.

Der Verlag behält sich fur alle Beiträge das Recht der Vervielfältigung, Verbreitung und Übersetzung vor.

Bezugsmöglichkeiten: Bestellungen sind zu richten

— **in der DDR** an den Postzeitungsvertrieb; an eine Buchhandlung oder an den Akademie-Verlag, DDR-108 Berlin, Leipziger Straße 3—4

— **im sozialistischen Ausland** an eine Buchhandlung für fremdsprachige Literatur oder an den zuständigen Postzeitungsvertrieb

— **in der BRD und Westberlin** an eine Buchhandlung oder an die Auslieferungsstelle KUNST UND WISSEN, Erich Bieber, 7 Stuttgart 1, Wilhelmstraße 4—6

— **in Österreich** an den Globus-Buchvertrieb, 1201 Wien, Höchstädtplatz 3

— **im übrigen Ausland** an den Internationalen Buch- und Zeitschriftenhandel; den Buchexport, Volkseigener Außenhandelsbetrieb der Deutschen Demokratischen Republik, DDR-701 Leipzig, Postfach 160, oder an den Akademie-Verlag, DDR-108 Berlin, Leipziger Straße 3—4

ZEITSCHRIFT FÜR ANGEWANDTE MATHEMATIK UND MECHANIK.
Herausgeber und Chefredakteur: Prof. Dr. Günter Schmidt (redaktionelle Mitarbeit: Dr. W. Heinrich und Dr. H. Weinert), Zentralinstitut für Mathematik und Mechanik der Akademie der Wissenschaften der DDR. Verlag: Akademie-Verlag, DDR-108 Berlin, Leipziger Straße 3—4; Fernruf: 2200441; Telex-Nr.: 114420; Postscheckkonto: Berlin 350/21. Bank: Staatsbank der DDR, Berlin, Kto.-Nr.: 6836-26-20712. Anschrift der Reaktion: Zentralinstitut für Mathematik und Mechanik der Akademie der Wissenschaften, DDR-108 Berlin, Mohrenstraße 39; Fernruf: 2000561.

Veroffentlicht unter der Lizenznummer 1282 des Presseamtes beim Vorsitzenden des Ministerrates der Deutschen Demokratischen Republik. Gesamtherstellung: VEB Druckerei „Thomas Müntzer", DDR-582 Bad Langensalza. Erscheinungsweise: Die Zeitschrift für Angewandte Mathematik und Mechanik erscheint monatlich. Die 12 Hefte eines Jahres einschließlich Tagungsheft bilden einen Band. Bezugspreis je Band 300,— M zuzuglich Versandspesen (Preis für die DDR 216,—M). Bezugspreis je Heft 25,— M (Preis für die DDR 18,—M). Bestellnummer dieses Heftes: 1009/56/11.

ZAMM 56, 449–452 (1976)

S. T. ARIARATNAM / D. S. F. TAM

Parametric Random Excitation of a Damped Mathieu Oscillator

Die Wirkung einer Zufallsparametererregung auf die Momentenstabilität eines gedämpften Mathieuschen Schwingers wird untersucht. Wenn die harmonische Erregung nahe bei der doppelten Eigenfrequenz des Schwingers liegt, werden Bedingungen für die Stabilität der ersten und zweiten Momente der Antwort erhalten. Es ergibt sich, daß die Anwesenheit der zusätzlichen Zufallsparametererregung entweder eine stabilisierende oder eine destabilisierende Wirkung haben kann, abhängig von den Werten bestimmter Parameter der Zufallserregung. Insbesondere wird gezeigt, daß Parametererregung durch Weißes Rauschen keinen Einfluß auf die ersten Momente der Antwort, aber einen destabilisierenden Einfluß auf die zweiten Momente hat. Für den Fall eines exponentiell korrelierten Rauschens werden die Instabilitätsbereiche graphisch dargestellt.

The effect of parametric random excitation on the moment stability of a damped Mathieu oscillator is investigated. Conditions for stability of the first and second moments of the response are obtained when the harmonic excitation lies in the neighbourhood of twice the natural frequency of the oscillator. It is found that the presence of the additional parametric random excitation can cause either a stabilizing or a destabilizing effect depending on the values of certain parameters of the random excitation. In particular, it is shown that parametric excitation by white noise has no effect on the first moments of the response but has a destabilizing effect on the second moments. For the case of exponentially correlated noise, regions of instability are shown in graphical form.

Исследуется влияние случайного параметрического возбуждения на устойчивость моментов колебателя Матье с затуханием. Если гармоническое возбуждение лежит вблизи двойной собественной частоты колебателя, получаются условия для устойчивости первого и второго моментов выхода. Оказывается, что наличие дополнительного случайного возбуждения может иметь стабилизирующее или дестабилизирующее воздействие в зависимости от величин определённых параметров случайного возбуждения. В частности, показывается, что параметрическое возбуждение белым шумом не влияет на первые моменты выхода, но оказывает дестабилизирующее воздействие на вторые моменты. Для случая экспоненциально коррелирующего шума области неустойчивости представляются графически.

1. Introduction

It is well-known that regions of instability for a MATHIEU oscillator are found when the ratio of the frequency of the harmonic excitation to twice the natural frequency of the oscillator is in the neighbourhood of the values $1/p$, where p is a positive integer. The most important of these corresponds to $p = 1$ and is known as the region of fundamental resonance. When the harmonic excitation is replaced by a stationary stochastic excitation, conditions for stochastic stability have been obtained by STRATONOVICH and ROMANOVSKII [1], WEIDENHAMMER [2] and GRAEFE [3] who showed that, to a first approximation, stability depends only on the value of the excitation spectrum in the neighbourhood of twice the system natural frequency. However, in some applications, the excitation may consist of a harmonic term on which a stochastic fluctuation is superposed. The almost sure (or sample) stability of a linear oscillator subjected to such a parametric excitation has been studied by STRATONOVICH and ROMANOVSKII [4]*. This paper investigates the moment stability of the same system when the frequency of the harmonic excitation lies in the region of fundamental parametric resonance. Conditions, valid in the first approximation, are derived for stability of the first and second moments and are found to depend on the random excitation only through its spectral values at zero frequency and at twice the natural frequency of the oscillator. Regions of instability are shown in graphical form for the case of random excitation having an exponential correlation function.

2. Formulation

The damped MATHIEU oscillator subjected to an additional parametric random excitation is described by a stochastic differential equation of the form

$$\ddot{x} + 2\varepsilon\zeta\omega_0\dot{x} + \omega_0^2\,[1 + \varepsilon h \sin \nu t + \varepsilon^{1/2} f(t)]\, x = 0\,, \tag{1}$$

where ζ, ω_0 represent, respectively, the damping ratio and the undamped natural frequency of the oscillator, h, ν denote the amplitude and frequency of the harmonic excitation and $f(t)$ is a stationary stochastic process with zero mean value; $\varepsilon \ll 1$ is a small parameter.

Considering the case of fundamental parametric resonance, i.e. when $\nu \cong 2\omega_0$, and setting

$$\omega_0^2 = \nu^2/4 + \varepsilon\Delta\,, \tag{2}$$

where $\varepsilon\Delta$ denotes the amount of detuning, equation (1) may be rewritten as

$$\ddot{x} + \frac{\nu^2}{4}\,x = -\,\varepsilon\,[2\zeta\omega_0\dot{x} + (\Delta + \omega_0^2 h \sin \nu t)\, x] - \varepsilon^{1/2}\omega_0^2\, f(t)\, x\,. \tag{3}$$

*) The kind of stability considered in reference [4], though not explicitly stated, can be shown to be the same as almost sure stability.

Transforming to new variables z_1, z_2 by the relations

$$x = z_1 \cos (\nu t/2) + z_2 \sin (\nu t/2)\,, \qquad \dot{x} = -\tfrac{1}{2} z_1 \nu \sin (\nu t/2) + \tfrac{1}{2} z_2 \nu \cos (\nu t/2)$$

equation (3) may be replaced by the pair of first-order equations

$$\dot{z}_1 = -\varepsilon\left[2\zeta\omega_0\left(z_1 \sin^2\frac{\nu}{2}t - \frac{1}{2} z_2 \sin \nu t\right) + \left(\frac{2\Delta}{\nu} + \omega_0 h \sin \nu t\right)\left(\frac{1}{2} z_1 \sin \nu t + z_2 \sin^2\frac{\nu}{2} t\right)\right] + $$
$$+ \varepsilon^{1/2}\omega_0 f(t)\left(\frac{1}{2} z_1 \sin \nu t + z_2 \sin^2 \frac{\nu}{2} t\right), \tag{4a}$$

$$\dot{z}_2 = \varepsilon\left[2\zeta\omega_0\left(\frac{1}{2} z_1 \sin \nu t - z_2 \cos^2\frac{\nu}{2} t\right) - \left(\frac{2\Delta}{\nu} + \omega_0 h \sin \nu t\right)\left(z_1 \cos^2\frac{\nu}{2} t + \frac{1}{2} z_2 \sin \nu t\right)\right] - $$
$$- \varepsilon^{1/2}\omega_0 f(t)\left(z_1 \cos^2\frac{\nu}{2} t + \frac{1}{2} z_2 \sin \nu t\right). \tag{4b}$$

If the relaxation time of the z_1, z_2 processes is much larger than the correlation time of the stochastic process $f(t)$, the solutions of equations (4) may be approximated over a time interval of order $O(\varepsilon^{-1})$ by a Markov vector process which is continuous with probability one. The procedure for this approximation has been given by Stratonovich [5] and by Khasminskii [6], and when applied to equations (4), leads to the following homogeneous Itô equations:

$$\mathrm{d}z_1 = \varepsilon\left[\omega_0\left(\frac{h}{4} - \zeta^*\right) z_1 + \left(\frac{\Delta}{\nu} - a\right) z_2\right] \mathrm{d}t + \varepsilon^{1/2} \left[\sigma_{11}(z)\, \mathrm{d}w_1 + \sigma_{12}(z)\, \mathrm{d}w_2\right], \tag{5a}$$

$$\mathrm{d}z_2 = \varepsilon\left[-\left(\frac{\Delta}{\nu} - a\right) z_1 - \omega_0\left(\frac{h}{4} + \zeta^*\right) z_2\right] \mathrm{d}t + \varepsilon^{1/2} \left[\sigma_{21}(z)\, \mathrm{d}w_1 + \sigma_{22}(z)\, \mathrm{d}w_2\right], \tag{5b}$$

where $w_1(t)$, $w_2(t)$ are independent Wiener processes of unit intensity and

$$\zeta^* = \zeta + \frac{\omega_0}{8}[S(0) - S(2\omega_0)]\,, \quad a = \frac{\omega_0^2}{8}\Psi(2\omega_0)\,, \quad [\sigma\sigma']_{11} = \frac{\omega_0^2}{8} S(2\omega_0)\, z_1^2 + \frac{\omega_0^2}{8}[2S(0) + S(2\omega_0)]\, z_2^2\,,$$

$$[\sigma\sigma']_{22} = \frac{\omega_0^2}{8}[2S(0) + S(2\omega_0)]\, z_1^2 + \frac{\omega_0^2}{8} S(2\omega_0)\, z_2^2\,, \qquad [\sigma\sigma']_{12} = [\sigma\sigma']_{21} = -\frac{\omega_0^2}{4} S(0)\, z_1 z_2\,,$$

$$S(\omega) = 2\int_0^\infty \langle f(t)\, f(t+\tau)\rangle \cos \omega\tau\, \mathrm{d}\tau\,, \qquad \Psi(\omega) = 2\int_0^\infty \langle f(t)\, f(t+\tau)\rangle \sin \omega\tau\, \mathrm{d}\tau;$$

the angular brackets denote the expectation or the ensemble average.

3. Stability Conditions

The differential equations governing the mean values of z_1, z_2 can be obtained by taking the expectation of both sides of equations (5). Thus

$$\frac{\mathrm{d}}{\mathrm{d}t}\begin{bmatrix}\langle z_1\rangle \\ \langle z_2\rangle\end{bmatrix} = \varepsilon \begin{bmatrix} \omega_0\left(\dfrac{h}{4} - \zeta^*\right) & \dfrac{\Delta}{\nu} - a \\ -\left(\dfrac{\Delta}{\nu} - a\right) & -\omega_0\left(\dfrac{h}{4} + \zeta^*\right)\end{bmatrix} \begin{bmatrix}\langle z_1\rangle \\ \langle z_2\rangle\end{bmatrix}. \tag{6}$$

Hence, the conditions for stability in the first moments are

$$\zeta^* > 0, \qquad \omega_0^2\zeta^{*2} + \left(\frac{\Delta}{\nu} - a\right)^2 > \frac{\omega_0^2 h^2}{16} \tag{7}$$

which are equivalent to

$$\zeta > \frac{\omega_0}{8}[S(2\omega_0) - S(0)]\,, \qquad \varepsilon h < 4\left[\left\{1 - \frac{\nu}{2\omega_0} - \varepsilon\frac{\omega_0}{8}\Psi(2\omega_0)\right\}^2 + \varepsilon^2\left\{\zeta - \frac{\omega_0}{8}[S(2\omega_0) - S(0)]\right\}^2\right]^{1/2}.$$

(8a) (8b)

In the case of white noise excitation, $S(\omega) = S_0 =$ constant, $\Psi(\omega) = 0$ and the conditions (8) reduce to

$$\zeta > 0\,, \qquad \varepsilon h < 4\left[\left(1 - \frac{\nu}{2\omega_0}\right)^2 + \varepsilon^2\zeta^2\right]^{1/2}. \tag{9}$$

It may be remarked that the first condition is the same as that for an oscillator without the sinusoidal excitation [7], while the second is identical to the stability condition in the absence of the stochastic excitation.

The differential equations satisfied by the second moments can be found either from the FOKKER-PLANCK equation associated with equations (5) (see, e.g. [7]) or by applying ITÔ's differential rule [8] to the quantities z_1^2, $z_1 z_2$, z_2^2. Either procedure leads to

$$\frac{\mathrm{d}}{\mathrm{d}t}\begin{bmatrix}\langle z_1^2\rangle\\ \langle z_1 z_2\rangle\\ \langle z_2^2\rangle\end{bmatrix} = 2\varepsilon\begin{bmatrix} d-\omega_0(\zeta-3b+c) & \frac{\Delta}{\nu}-a & c+b\\ -\frac{1}{2}\left(\frac{\Delta}{\nu}-a\right) & -\omega_0(\zeta-2b+2c) & \frac{1}{2}\left(\frac{\Delta}{\nu}-a\right)\\ c+b & -\left(\frac{\Delta}{\nu}-a\right) & -d-\omega_0(\zeta-3b+c)\end{bmatrix}\begin{bmatrix}\langle z_1^2\rangle\\ \langle z_1 z_2\rangle\\ \langle z_2^2\rangle\end{bmatrix} \tag{10}$$

where

$$b=\frac{\omega_0}{16}S(2\omega_0)\,,\qquad c=\frac{\omega_0}{8}S(0)\,,\qquad d=\frac{\omega_0 h}{4}.$$

From equations (10), the conditions for stability in the second moments (i.e. mean square stability) are found to be

$$\zeta>4b\,,\qquad \frac{\omega_0^2 h^2}{16}<\left[\left(\frac{\Delta}{\nu}-a\right)^2+\omega_0^2(\zeta+2c-2b)^2\right]\frac{\zeta-4b}{\zeta+2c-2b} \tag{11a), (11b}$$

which are equivalent to

$$\zeta>\frac{\omega_0}{4}S(2\omega_0)\,, \tag{12a}$$

$$\varepsilon h<4\left[\left\{1-\frac{\nu}{2\omega_0}-\varepsilon\frac{\omega_0}{8}\Psi(2\omega_0)\right\}^2+\varepsilon^2\left\{\zeta+\frac{\omega_0}{4}S(0)-\frac{\omega_0}{8}S(2\omega_0)\right\}^2\right]^{1/2}\left[\frac{8\zeta-2\omega_0 S(2\omega_0)}{8\zeta+2\omega_0 S(0)-\omega_0 S(2\omega_0)}\right]^{1/2}. \tag{12b}$$

In the case of white noise excitation of intensity S_0, the conditions reduce to

$$\zeta>\frac{\omega_0 S_0}{4},\qquad \varepsilon h<4\left[\left(1-\frac{\nu}{2\omega_0}\right)^2+\varepsilon^2\left(\zeta+\frac{\omega_0 S_0}{8}\right)^2\right]^{1/2}\left[\frac{8\zeta-2\omega_0 S_0}{8\zeta+\omega_0 S_0}\right]^{1/2}. \tag{13a), (13b}$$

It can be easily seen that the region of instability is widened by the presence of the additional white noise excitation. As an example of non-white random excitation, suppose that $f(t)$ is exponentially correlated, so that

$$\langle f(t)\,f(t+\tau)\rangle=\sigma^2\exp\left(-\alpha|\tau|\right).$$

Then

$$S(\omega)=\frac{2\sigma^2\alpha}{\alpha^2+\omega^2},\qquad \Psi(\omega)=\frac{2\sigma^2\omega}{\alpha^2+\omega^2}$$

and the stability conditions (8)' and (12) become

$$\zeta>-\frac{\sigma^2}{8k(1+k^2)},\qquad \varepsilon h<4\left[\left\{1-\frac{\nu}{2\omega_0}-\frac{\varepsilon\sigma^2}{8(1+k^2)}\right\}^2+\varepsilon^2\left\{\zeta+\frac{\sigma^2}{8k(1+k^2)}\right\}^2\right]^{1/2} \tag{14a), (14b}$$

for the first moments, and

$$\zeta>\frac{\sigma^2 k}{4(1+k^2)}, \tag{15a}$$

$$\varepsilon h<4\left[\left\{1-\frac{\nu}{2\omega_0}-\frac{\varepsilon\sigma^2}{8(1+k^2)}\right\}^2+\varepsilon^2\left\{\zeta+\frac{\sigma^2(2+k^2)}{8k(1+k^2)}\right\}^2\right]^{1/2}\left[\frac{8k(1+k^2)\zeta-2\sigma^2k^2}{8k(1+k^2)\zeta+\sigma^2(2+k^2)}\right]^{1/2} \tag{15b}$$

for the second moments, where $k=\alpha/(2\omega_0)$.

For fixed values of the parameters ζ, σ^2, ε, the instability regions in the $(h, \nu/2\omega_0)$-plane given by (14) and (15) are shown in Figs. 1 and 2 for different values of k. It may be seen that, for a given value of the damping coefficient ζ, by a suitable choice of σ^2 and k, it is possible to stabilize an unstable MATHIEU oscillator.

The relaxation time of the system given by equations (5) is $O(\varepsilon^{-1})$, while the correlation time of the excitation $f(t)$ is $O(\alpha^{-1})$. Hence the results obtained above are valid provided $\alpha\gg\varepsilon$, or $k\omega_0\gg\varepsilon$.

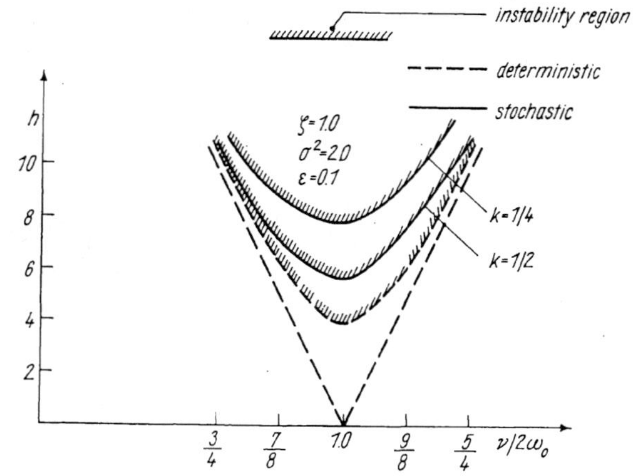

Fig. 1. Instability Regions for the First Moments

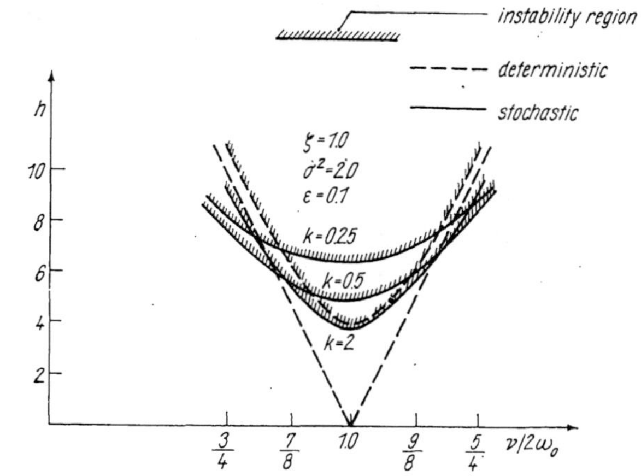

Fig. 2. Instability Regions for the Second Moments

Acknowledgements

The research of this paper was supported by the National Research Council of Canada through Grant No. A 1815.

References

1 Stratonovich, R. L., and Romanovskii, Yu. M., Non-Linear Transformations of Random Processes, Ed.: P. I. Kuznetsov, R. L. Stratonovich and V. I. Tikhonov, Pergamon p. 322 (1965).
2 Weidenhammer, F., Stabilitätsbedingungen für Schwinger mit zufälligen Parametererregungen, Ing.-Arch. **33**, 404 (1964).
3 Graefe, P. W. U., Stability of a Linear Second Order System under Random Parametric Excitation, Ing.-Arch. **35**, 276 (1966).
4 Stratonovich, R. L., and Romanovskii, Yu. M., ibid. p. 327.
5 Stratonovich, R. L., Topics in the Theory of Random Noise. Vol. **1**, Gordon and Breach, New York 89 (1963).
6 Khasminskii, R. Z., A Limit Theorem for the Solutions of Differential Equations with Random Right-Hand Sides, Theory Prob. Appl. **11**, 390 (1966).
7 Ariaratnam, S. T., Dynamic Stability of Structures, Ed.: G. Herrmann, Pergamon, New York, 255 (1967).
8 Itô, K., On Stochastic Differential Equations, Memoirs of the American Mathematical Society **4**, 51 (1951).

Eingereicht am 20. 9. 1975

Anschriften: Professor Dr. S. T. Ariaratnam, Dr. D. S. F. Tam, Solid Mechanics Division, University of Waterloo, Waterloo, Ontario, N2L 3G1, Canada

ZAMM 56, 453 – 459 (1976)

K. GOPALSAMY

On a Class of Linear Systems with Random Coefficients

In der Arbeit wird die mittlere quadratische Stabilität von zufallsgestörten linearen Systemen betrachtet. Anstelle einer Betrachtung von partiellen Differentialgleichungen werden abstrakte Differentialgleichungen untersucht. Indem die Störungstheorie einer Operatorenhalbgruppe verwendet wird, können hinreichende Bedingungen für Zufallsstörkoeffizienten erhalten werden, so daß die mittlere quadratische Abweichung des gestörten Systems vom ungestörten klein bleibt. Es wird eine Wahrscheinlichkeitsschranke dafür erhalten, daß das gestörte System asymptotisch stabil ist.

Mean-square stability of randomly perturbed linear systems is considered. Instead of considering partial differential equations, abstract differential equations are studied. Using the perturbation theory of a semi-group of operators, sufficient conditions on the perturbing random coefficient are obtained so that the mean-square deviation of the perturbed system response from that of the unperturbed will remain small. A probability bound is obtained for the perturbed system to be asymptotically stable.

Рассматривается средняя квадратная устойчивость случайно-возмущенной линейной системы. Вместо рассмотрения дифференциальных уравнений в частных произведениях изучаются абстрактные дифференциальные уравнения Применяя теорию возмущения полугруппы операторов получаются достаточные условия для возмущенных коэффициентов так что среднее крвадратичное отклонение возмущенной системы от невозмущенной остается мелко. Получается вероятностная граница для асимптотической устойчивости возмущенной системы.

1. Introduction

Stability of systems with random parametric excitation was considered by many authors in the literature. Those cases where the random excitation is related to a white noise were successfully investigated using the methods and the results of the theory of MARKOV processes. Investigation of the stability of randomly excited systems when non-white noise excitation is involved is difficult and hence most workers have considered only systems of particular type. Systems governed by partial differential equations under random excitations are not considered by many authors in the literature. A number of problems involving partial differential equations can be considered by the theory of semi-groups of operators. In the following we will use the perturbation theory of semi-groups of operators to obtain a sufficient condition for the meansquare stability of a class of randomly excited linear systems governed by a generator of a contraction semi-group of bounded linear operators. The analysis is conveniently done in a function space which we will introduce now.

Let (Ω, Σ, μ) be a probability measure space on which all random variables considered below are defined. Let J denote a finite time interval defined by $J = \{0 \leqq t \leqq T < \infty\}$ and let X be a BANACH space. Let $\{u(t), t \in J\}$ be a random process with values in an arbitrary space Y of random variables. The random process $\{u(t), t \in J\}$ can be considered to be a mapping $u(t, \omega)$ from $J \times \Omega$ into Y. The random variable $u(t) = u(t, \cdot)$ will be identified with an element of a BANACH space X which is generated by Y in the following sense. If Y is a BANACH space, we will take $X = L^p(Y)$ $(1 \leqq p < \infty)$, the space of equivalence classes of strongly measurable Y-valued functions f on (Ω, Σ, μ) for which

$$\int_\Omega \|f(\omega)\|^p \, d\mu < \infty \qquad f \in Y\,, \qquad 1 \leqq p < \infty\,.$$

The random process $\{u(t), t \in J\}$ is said to be *strongly differentiable* at t_0 with a *strong derivative* $\dot{u}(t_0)$ if

$$\lim_{h\to 0} \mathsf{E}\left[\left\|\frac{u(t_0+h)-u(t_0)}{h} - \dot{u}(t_0)\right\|\right] = 0\,.$$

If Y is the real line, then the above limit can be taken in the form

$$\lim_{h\to 0} \mathsf{E}\left[\left|\frac{u(t_0+h)-u(t_0)}{h} - \dot{u}(t_0)\right|^p\right] = 0$$

where $\mathsf{E}[\ ]$ denotes expectation. For $p = 2$, $\dot{u}(t_0)$ is known as the *mean square derivative*. In the following we take Y to be the real line.

We consider an *unperturbed* linear system

$$\frac{du}{dt} = Au + f(t)\,, \qquad 0 < t \leqq T\,, \tag{1.1}$$

and a randomly *perturbed* system

$$\frac{dv}{dt} = (A + B(t))\, v + f(t) + r(t)\,, \qquad 0 < t \leqq T\,. \tag{1.2}$$

It is reasonable to expect that if the random perturbations $B(\cdot)$ and $r(\cdot)$ are small in some sense, then the solutions of (1.2) will not differ much from those of (1.1).

A number of boundary value problems involving partial differential equations can be represented in the form of (1.1) and (1.2). The partial differential operator A is assumed to satisfy the following

Conditions: A is a closed linear operator the domain of definition $D(A)$ of which is dense in X and for some $M > 0, \alpha > 0$ we have

$$||(\lambda I - A)^{-n}|| \leqq M(\lambda + \alpha)^{-n}, \quad n = 1, 2, 3, \ldots \tag{1.3}$$

for all $\lambda > -\alpha$. It follows from the Feller-Miyadera-Phillips generation theorem (cf. Hille and Phillips [1], p. 360) that the operator A generates a *strongly continuous semi-group of contraction operators* $\{T(t), t \in J\}$ on X satisfying

$$||T(t)|| \leqq M\,e^{-\alpha t}, \qquad t \in J\,. \tag{1.4}$$

We assume furthermore that $\{B(t), t \in J\}$ is a family of continuously differentiable mappings from $J \times \Omega$ into the space of bounded linear endomorphisms on X; $f(t)$ and $r(t)$ $(t \in J)$ are continuously differentiable mappings from J and $J \times \Omega$ into X, respectively. $\{r(t), t \in J\}$ and $\{B(t), t \in J\}$ are assumed to be statistically independent.

2. Stability under random perturbations

There are a uumber of concepts of *stability* of random systems in the literature. For a detailed bibliography, terminology and definitions we refer to Kushner [2] or Kozin [3]. In the following we consider the behaviour of a linear system under a class of random perturbations.

Suppose the linear system

$$\frac{du}{dt} = Au\,, \qquad 0 < t \leqq T\,, \qquad u(0) = u_0\,, \quad u_0 \in D(A) \subset X\,, \tag{2.1}$$

goes under a random perturbation of A into

$$\frac{dv}{dt} = (A + B(t))\,v\,, \qquad 0 < t \leqq T\,, \qquad v(0) = u_0\,, \quad u_0 \in D(A) \subset X\,, \tag{2.2}$$

where A and $B(\cdot)$ are as described earlier. Let

$$\psi(\omega) = \sup_{t \in J} |v(t) - u(t)|\,, \qquad \omega \in \Omega\,. \tag{2.3}$$

We will say that the system (2.1) is *stable under the random perturbation* $B(t)$ if $\mathsf{E}[\psi^2(\omega)]$ can be made arbitrarily small by taking $B(\cdot)$ small in the sense described below.

Theorem 1: *Let δ be an arbitrary positive constant. Then*

$$\mathsf{E}\left[\exp\left(2M \int_0^T ||B(s)||\,ds\right)\right] < 1 + \delta\,, \qquad t \in J\,, \tag{2.4}$$

$$\Rightarrow \mathsf{E}[\psi^2(\omega)] \leqq \varkappa\delta \tag{2.5}$$

for some $\varkappa > 0$.

Proof. It is known that [4],

$$v(t) - u(t) = \sum_{n=1}^{\infty} S_n(t)\,u_0\,,$$

where

$$S_1(t) = \int_0^t T(t-s)\,B(s)\,T(s)\,ds\,, \qquad t \in J\,, \qquad S_n(t) = \int_0^t T(t-s)\,B(s)\,S_{n-1}(s)\,ds\,, \qquad n > 1\,,$$

$\{T(t), t \in J\}$ being the semigroup of contraction operators generated by the operator A. As indicated earlier we know that $||T(t)|| \leqq M\,e^{-\alpha t}$, $\alpha > 0$, $t \in J$. Hence

$$||S_1(t)|| \leqq M^2\,e^{-\alpha t} \int_0^t ||B(s)||\,ds\,, \qquad t \in J\,, \qquad ||S_2(t)|| \leqq M^3\,e^{-\alpha t}\left(\int_0^t ||B(s)||\,ds\right)^2 \Big/ 2!\,,$$

$$||S_n(t)|| \leqq M^{n+1}\,e^{-\alpha t}\left[\int_0^t ||B(s)||\,ds\right]^n \Big/ n!\,, \qquad |v(t) - u(t)| \leqq M\,e^{-\alpha t}\left[\exp\left\{M \int_0^T ||B(s)||\,ds\right\} - 1\right]|u_0|\,,$$

$$\psi(\omega) \leqq M\left[\exp\left\{M \int_0^T ||B(s)||\,ds\right\} - 1\right]|u_0|\,,$$

$$\psi^2(\omega) \leqq M^2\left[\exp\left(2M \int_0^T ||B(s)||\,ds\right) - 2\exp\left(M \int_0^T ||B(s)||\,ds\right) + 1\right]|u_0|^2\,,$$

$$\leqq M^2\left[\exp\left(2M \int_0^T ||B(s)||\,ds\right) - 1\right]|u_0|^2\,,$$

$$\mathsf{E}[\psi^2(\omega)] \leqq M^2\,|u_0|^2\,\delta\,.$$

Thus it follows that the system (2.1) is stable under the perturbation $B(t)$ of the operator A. An example of the operator A satisfying the condition (1.3) is shown in [5], [6].

Theorem 2: *Suppose now that the system*

$$\frac{du}{dt} = Au + f(t)\,, \qquad 0 < t \leqq T\,, \qquad u(0) = 0\,,$$

goes under a perturbation into

$$\frac{dv}{dt} = \bigl(A + B(t)\bigr)\, v + f(t) + r(t)\,, \qquad 0 < t \leqq T\,, \qquad v(0) = 0\,.$$

Assume that the perturbations $r(t)$ and $B(t)$ are small in the sense that for an arbitrary $\delta > 0$ we have

$$\int_0^T \mathsf{E}\, |r(t)|^2\, dt < \delta\,, \qquad \mathsf{E}\left[\exp\left(2M \int_0^T ||B(s)||\, ds\right)\right] < 1 + \delta\,.$$

Let $\int_0^T |f(t)|\, dt = N_1$. If $\psi(\omega) = \sup_{t \in J} |v(t) - u(t)|$, then we have

$$\mathsf{E}[\psi^2(\omega)] \leqq 2M^2\, (N_1 + 2)\, \delta$$

(δ being a perturbation scale we assume $0 < \delta < 1$).

Proof. From the perturbation theory of semi-groups [4],

$$u(t) = \int_0^t T(t-s)\, f(s)\, ds\,, \qquad t \in J\,, \qquad v(t) = \sum_{n=0}^{\infty} [W_n^{(1)}(t) + W_n^{(2)}(t)]\,,$$

where

$$W_0^{(1)}(t) = \int_0^t T(t-s)\, f(s)\, ds\,, \qquad W_n^{(1)}(t) = \int_0^t T(t-s)\, B(s)\, W_{n-1}^{(1)}(s)\, ds\,, \qquad n \geqq 1\,,$$

$$W_0^{(2)}(t) = \int_0^t T(t-s)\, r(s)\, ds\,, \qquad W_n^{(2)}(t) = \int_0^t T(t-s)\, B(s)\, W_{n-1}^{(2)}(s)\, ds\,, \qquad n \geqq 1\,,$$

$$v(t) - u(t) = \sum_{n=1}^{\infty} W_n^{(1)}(t) + \sum_{n=0}^{\infty} W_n^{(2)}(t)\,, \qquad |W_0^{(1)}(t)| \leqq MN_1\,, \qquad t \in J\,,$$

$$|W_1^{(1)}(t)| \leqq N_1 M^2 \int_0^t ||B(s)||\, ds\,, \qquad |W_n^{(1)}(t)| \leqq N_1 M^{n+1} \left(\int_0^t ||B(s)||\, ds\right)^n /n!\,, \qquad n = 0, 1, 2, \ldots\,,$$

$$|W_0^{(2)}(t)| \leqq M \int_0^T |r(s)|\, ds = MN_2(\omega) \quad \text{(say)} \quad \omega \in \Omega\,, \qquad t \in J\,.$$

$$|W_1^{(2)}(t)| \leqq M^2 N_2(\omega) \int_0^t ||B(s)||\, ds\,, \qquad t \in J\,,$$

$$|W_n^{(2)}(t)| \leqq M^{n+1} N_2(\omega) \left(\int_0^t ||B(s)||\, ds\right)^n /n!\,, \qquad n = 0, 1, 2, \ldots\,,$$

$$|v(t) - u(t)| \leqq MN_1 \left[\exp\left(M \int_0^t ||B(s)||\, ds\right) - 1\right] + MN_2(\omega) \left[\exp\left(M \int_0^t ||B(s)||\, ds\right)\right]$$

$$\leqq MN_1 \left[\exp\left(M \int_0^T ||B(s)||\, ds\right) - 1\right] + MN_2(\omega) \left[\exp\left(M \int_0^T ||B(s)||\, ds\right)\right].$$

Using $(x + y)^2 \leqq 2(x^2 + y^2)$,

$$\psi^2(\omega) \leqq 2M^2 N_1^2 \left[\exp\left(2M \int_0^T ||B(s)||\, ds\right) - 2\exp\left(M \int_0^T ||B(s)||\, ds\right) + 1\right] +$$

$$+ 2M^2 N_2^2(\omega) \left[\exp\left(2M \int_0^T ||B(s)||\, ds\right)\right]$$

$$\leqq 2M^2 N_1^2 \left[\exp\left(2M \int_0^T ||B(s)||\, ds\right) - 1\right] + 2M_2 N_2^2(\omega) \left[\exp\left(2M \int_0^T ||B(s)||\, ds\right)\right],$$

$$\mathsf{E}[\psi^2(\omega)] \leqq 2M^2 N_1^2 \delta + 2M^2\, (1 + \delta)\, \mathsf{E}[N_2^2(\omega)] \leqq 2M^2 N_1^2 \delta + 2M^2\, (1 + \delta)\, \delta \leqq 2M^2\, (N_1^2 + 2)\, \delta$$

$$(\because \delta < 1)\,.$$

3. A probability bound for asymptotic stability

We will now consider the behaviour of random linear systems over an infinite time interval. Let $R^+ = \{t\colon 0 \leqq t < \infty\}$ and consider the space of all continuous functions defined on R^+ with values in the space Y of scalar valued random variables.

Definition 1: The space C_g denoted by

$$C_g = C_g(R^+, L_2(\Omega, \Sigma, \mu))$$

is the Banach *space of continuous functions* $u(\cdot)$ *defined on* R^+ *with values in* Y such that

$$\Big\{\int_\Omega |u(t)|^2 \,\mathrm{d}\mu\Big\}^{1/2} \leqq \gamma g(t)\,, \qquad t \in R^+\,,$$

where γ is a positive constant and $g(t)$ is a given positive valued continuous function on R^+.

Definition 2: A random function $\{u(t), t \in R^+\}$ is said to be *asymptotically stable about the null function with probability* ϱ_u if

$$\varrho_u = \mathsf{P}\left\{\lim_{t\to\infty} |u(t)| = 0\right\}.$$

If $\varrho_u = 1$, *almost sure stability* is obtained. In the following the Banach space considered is a C_g for some γ and $g(t)$.

We first illustrate the destabilizable character of random perturbations. Suppose the system

$$\frac{\mathrm{d}u}{\mathrm{d}t} = Au\,, \qquad t > 0\,, \quad u(0) = u_0$$

is stable in the sense that

$$\lim_{t\to\infty} |u(t)|^2 = \lim_{t\to\infty} |T(t)\, u_0|^2 < \infty$$

and consider a perturbed system

$$\frac{\mathrm{d}v}{\mathrm{d}t} = \big(A + B(\omega)\big)\, v\,, \qquad t > 0\,, \qquad \omega \in \Omega\,, \qquad v(0) = u_0\,,$$

where $\{B(\omega), \omega \in \Omega\}$ is a normal random variable with zero mean and variance σ^2. It is found that

$$v(t) - u(t) = [\mathrm{e}^{B(\omega)t} - 1]\, T(t)\, u_0\,, \qquad [v(t) - u(t)]^2 = [\mathrm{e}^{2B(\omega)t} - 2\mathrm{e}^{B(\omega)t} + 1]\, |u(t)|^2\,,$$

$$\mathsf{E}[v(t) - u(t)]^2 = [\mathrm{e}^{2\sigma^2t^2} - 2\mathrm{e}^{(t^2/2)\sigma^2} + 1]\, |u(t)|^2\,.$$

For large values of t,

$$\mathsf{E}[v(t) - u(t)]^2 \geqq \mathrm{e}^{(t^2/2)\sigma^2}\, (\mathrm{e}^{(3/2)t^2\sigma^2} - 2)\, |u(t)|^2 \to \infty \quad \text{as} \quad t \to \infty\,,$$

since $u(t)$ is stable. (This happens as long as $u(t)$ is a function of exponential order.) This happens even if $|u(t)| \to 0$ as $t \to \infty$. For example consider the simplest one dimensional system $\mathrm{d}u/\mathrm{d}t = -ru$ where r is a non-negative constant; here $u(t) = \mathrm{e}^{-rt}u_0$. In this case we have

$$\mathsf{E}[v(t) - u(t)]^2 \geqq \mathrm{e}^{\frac{t^2}{2}\sigma^2}\left(\mathrm{e}^{\frac{3}{2}t^2\sigma^2} - 2\right)\mathrm{e}^{-2rt}u_0^2 = \mathrm{e}^{t\frac{\sigma^2}{2}\left(t - \frac{4r}{\sigma^2}\right)}\left(\mathrm{e}^{\frac{3}{2}t^2\sigma^2} - 2\right)u_0^2 \to \infty$$

as $t \to \infty$ though $u(t) \to 0$ as $t \to \infty$.

Theorem: *Let*

$$\frac{\mathrm{d}u}{\mathrm{d}t} = \big(A + B(t)\big)\, u\,, \qquad 0 < t < \infty\,, \qquad u(0) = u_0\,, \qquad u_0 \in D(A)\,.$$

If $\{B(t), t \in R^+\}$ *is stationary and ergodic in the sense that for sufficiently large values of* t, *we have*

$$\frac{1}{t}\int_0^t ||B(s)||\, \mathrm{d}s = \mathsf{E}[||B(t)||] = a \text{ constant} = \beta \text{ (say)}$$

with probability one, then

$$|u(t)| \leqq M \exp\{-(\alpha - M\beta)\, t\}$$

for large values of t with probability one.

Proof: From

$$u(t) = T(t)\, u_0 + \int_0^t T(t-s)\, B(s)\, u(s)\, \mathrm{d}s$$

we have

$$|u(t)| \leqq M\,e^{-\alpha t}\,|u_0| + M \int_0^t e^{-\alpha(t-s)}\,||B(s)||\,|u(s)|\,ds\,.$$

Applying GRONWALL's inequality we have

$$|u(t)| \leqq M|u_0| \exp\left\{-\alpha t + M \int_0^t ||B(s)||\,ds\right\} \leqq M|u_0| \exp\left\{-\left(\alpha - \frac{M}{t}\int_0^t ||B(s)||\,ds\right)t\right\}, \qquad t \in R^+\,,$$

from which the result follows.

Corollary. *If*

$$\frac{du}{dt} = \big(A + B(t)\big)\,u\,, \qquad t > 0, \qquad u(0) = u_0\,,$$

then

$$\varrho_u \geqq \mathsf{P}\left\{\frac{1}{t}\int_0^t ||B(s)||\,ds < \frac{\alpha}{M}\right\}.$$

The following approach provides a one parameter family of deterministic upper bounds for the solution $u(t)$ of $du/dt = \big(A + B(t)\big)\,u$.

If we write $B(t) = B^*(t) + \mathsf{E}[B(t)]$ where $B^*(t) = B(t) - \mathsf{E}[B(t)]$ then from

$$\frac{du}{dt} = \big(A + B^*(t) + \mathsf{E}[B(t)]\big)\,u\,, \qquad t > 0\,, \qquad u(0) = u_0\,,$$

we get

$$u(t) = T(t)\,u_0 + \int_0^t T(t-s)\,\big(B^*(s) + \mathsf{E}[B(s)]\big)\,u(s)\,ds\,.$$

Hence

$$|u(t)| \leqq M\,e^{-\alpha t}\,|u_0| + \int_0^t M e^{-\alpha(t-s)}\,\{||B^*(s)|| + \mathsf{E}[||B(s)||]\}\,|u(s)|\,ds \leqq$$

$$\leqq M|u_0| \exp\left\{M \int_0^t \big(||B^*(s)|| + \mathsf{E}[||B(s)||]\big)\,ds - \alpha t\right\} \leqq M(|u_0| \exp\left\{M \int_0^t ||B^*(s)||\,ds - \alpha t + M\beta t\right\}$$

for large values of t.

By taking CHEBYSHEV's *inequality* in the form

$$\mathsf{P}\left\{\int_0^t ||B^*(s)||\,ds \leqq \delta\left(\int_0^t\int_0^t \mathsf{E}[||B^*(p)||\,||B^*(q)||]\,dp\,dq\right)^{1/2}\right\} \geqq 1 - \delta^{-2} \qquad \text{for any } \delta > 1$$

we get that

$$|u(t)| \leqq M|u_0| \exp\left\{-\,(\alpha - M\beta)\,t + \delta\left[\int_0^t\int_0^t \mathsf{E}[||B^*(p)||\,||B^*(q)||]\,dp\,dq\right]^{1/2}\right\},$$

with probability not less than $1 - \delta^{-2}$ for any $\delta > 1$. The case $\delta \leqq 1$ is of no interest. The following examples are illustrative. For applications to vibrating systems we refer to [7].

4. Examples

Example 1: The transverse displacements of a simply supported vibrating beam of length l are known to be governed by

$$EI\,\frac{\partial^4 u}{\partial x^4} + \varrho\,\frac{\partial^2 u}{\partial t^2} = f(x, t)\,, \qquad 0 < x < l\,, \qquad t > 0\,,$$

where

EI = bending stiffness of the beam, ϱ = linear mass density of the beam,
u = transverse displacement, f = transverse load per unit length.

Suppose under a perturbation the transverse displacements are governed by

$$EI\,\frac{\partial^4 u}{\partial x^4} + P(t)\,\frac{\partial^2 u}{\partial x^2} + \varrho\,\frac{\partial^2 u}{\partial t^2} = f(x, t)\,, \qquad 0 < x < l\,, \qquad t > 0\,,$$

where

$P(t)$ = random perturbation load intensity in the axial direction of the beam.

A physical situation where a perturbation of the above type arises is concerned with the stability of suspension bridges at times of earth tremors or earthquakes; during such instances there will be ground motions whose effects are transmitted to the motion of the bridge itself through the supporting towers.

We will show how the perturbed equation of motion can be reduced to the type of equation considered in the previous section.

Let us take the boundary and initial conditions to be

$$u|_{t=0} = 0\,, \qquad \left.\frac{\partial u}{\partial t}\right|_{t=0} = 0\,, \qquad u|_{x=0} = 0\,, \qquad u|_{x=l} = 0\,, \qquad \left.\frac{\partial^2 u}{\partial x^2}\right|_{x=0} = 0\,, \qquad \left.\frac{\partial^2 u}{\partial x^2}\right|_{x=l} = 0\,.$$

We assume that the independent variables are normalized so that $EI = 1$ and $\varrho = 1$. Writing an eigenfunction development

$$u(x,t) = \sum_{n=1}^{\infty} u_n(t) \sin\frac{n\pi x}{l}, \qquad f(x,t) = \sum_{n=1}^{\infty} f_n(t) \sin\frac{n\pi x}{l},$$

in the following, we consider only the nth mode of vibration. The unperturbed motion is governed by

$$\frac{\mathrm{d}Z}{\mathrm{d}t} = AZ + F(t)\,, \qquad Z(0) = 0\,,$$

where

$$Z = \begin{pmatrix} u_n \\ v_n \end{pmatrix}, \qquad \begin{aligned} \dot{u}_n &= \frac{n^2\pi^2}{l^2} v_n(t)\,, \\ \dot{v}_n &= -\frac{n^2\pi^2}{l^2} u_n(t)\,, \end{aligned}$$

$$A = \begin{pmatrix} 0 & \dfrac{n^2\pi^2}{l^2} \\ -\dfrac{n^2\pi^2}{l^2} & 0 \end{pmatrix}, \qquad F(t) = \begin{pmatrix} 0 \\ \dfrac{f_n(t)}{\dfrac{n^2\pi^2}{l^2}} \end{pmatrix},$$

i.e.

$$Z = \begin{pmatrix} u_n \\ v_n \end{pmatrix} = \int_0^t T(t-s) \begin{pmatrix} 0 \\ \dfrac{f_n(s)}{\dfrac{n^2\pi^2}{l^2}} \end{pmatrix} \mathrm{d}s\,, \qquad T(t) = \begin{pmatrix} \cos\dfrac{n^2\pi^2}{l^2}t & \sin\dfrac{n^2\pi^2}{l^2}t \\ -\sin\dfrac{n^2\pi^2}{l^2}t & \cos\dfrac{n^2\pi^2}{l^2}t \end{pmatrix}.$$

The perturbed motion is governed by

$$\frac{\mathrm{d}Z^*}{\mathrm{d}t} = (A + B(t))\,Z^* + F(t)\,, \qquad Z^*(0) = 0\,,$$

where

$$B(t) = \begin{pmatrix} 0 & 0 \\ P(t) & 0 \end{pmatrix},$$

i.e.

$$Z^*(t) = \sum_{j=0}^{\infty} S_j(t)\,,$$

$$S_0(t) = \int_0^t T(t-s) \begin{pmatrix} 0 \\ \dfrac{f_n(s)}{\dfrac{n^2\pi^2}{l^2}} \end{pmatrix} \mathrm{d}s\,,$$

$$S_j(t) = \int_0^t T(t-s)\, B(s)\, S_{j-1}(s)\, \mathrm{d}s\,, \qquad j \geqq 1\,.$$

An estimate of the deviation of the nth mode of the perturbed motion from that of the unperturbed motion is obtained as follows:

If

$$A = (a_{ij})\,, \qquad i, j = 1, 2, \dots, m\,,$$

then

$$||A|| = \sum_{j=1}^{m} \sum_{i=1}^{m} |a_{ij}| , \qquad ||Z^*(t) - Z(t)|| = \sum_{j=1}^{\infty} S_j(t)|| , \qquad ||T(t)|| \leqq 4 ,$$

$$||S_0(t)|| \leqq \frac{4}{\left(\frac{n^2\pi^2}{l^2}\right)} \int_0^t |f_n(s)| \, ds \leqq \frac{4}{\left(\frac{n\pi}{l}\right)^2} \int_0^T |f_n(s)| \, ds = \frac{4}{\left(\frac{n\pi}{l}\right)^2} N_1; \qquad N_1 = \int_0^T |f_n(s)| \, ds ,$$

$$||S_1(t)|| \leqq \frac{4N_1}{\left(\frac{n\pi}{l}\right)^2} \cdot 4 \int_0^t |P(s)| \, ds , \qquad ||S_2(t)|| \leqq \frac{4N_1}{\left(\frac{n\pi}{l}\right)} \frac{4^2 \left(\int_0^t |P(s)| \, ds\right)^2}{2} .$$

Hence

$$||Z^*(t) - Z(t)|| \leqq \frac{4N_1}{\left(\frac{n\pi}{l}\right)^2} \left[\exp\left(4 \int_0^t |P(s)| \, ds\right) - 1\right] \leqq \frac{4N_1}{\left(\frac{n\pi}{l}\right)^2} \left[\exp\left(4 \int_0^T |P(s)| \, ds\right) - 1\right],$$

$$\psi(\omega) = \sup_{0 \leqq t \leqq T} ||Z^*(t) - Z(t)|| \leqq \frac{4N_1}{\left(\frac{n\pi}{l}\right)^2} \left[\exp\left(4 \int_0^T |P(s)| \, ds\right) - 1\right],$$

$$\psi^2(\omega) \leqq \frac{16N_1^2}{\left(\frac{n^2\pi^2}{l^2}\right)^2} \left[\exp\left(8 \int_0^T |P(s)| \, ds\right) - 1\right],$$

$$\mathsf{E}[\psi^2(\omega)] \leqq \frac{16N_1^2}{\left(\frac{n^2\pi^2}{l^2}\right)} \delta \quad \text{if} \quad \mathsf{E}\left[\exp\left(\delta \int_0^T |P(s)| \, ds\right)\right] < 1 + \delta .$$

Example 2. Let us consider the random operator $B(t)$ to be a multiplication operator defined by

$$B(t)\, u(t) = \left(\int_0^t (-1)^{N(s)} \, ds\right) u(t)$$

where $N(t)$, $t \geqq 0$ is a POISSON process with parameter ν, $B(t)$ is the integral of a semi-random telegraph signal (for details see [8]). For a $B(t)$ of this type, if $u(t)$ is a solution of $du/dt = \big(A + B(t)\big)\, u$, $u(0) = u_0$ where A satisfies the conditions stated in the introduction then we have that for any $\delta > 1$, $t > 0$

$$||u(t)|| \leqq M \, ||u(0)|| \exp\left[-\alpha t + M\delta \left\{\frac{1}{2\nu^2}\left(e^{-2\nu t} - 1 + 2\nu t\right)\right\}^{1/2}\right]$$

with probability not less than $1 - \delta^{-2}$.

References

1 HILLE, E. and PHILLIPS, R. S., Functional Analysis and semi-groups, American Mathematical Society, Providence, R. I., 1957.
2 KUSHNER, H. J., Stochastic stability and control, Academic Press, 1967.
3 KOZIN, F., A survey of stability of stochastic systems in stochastic problems in control, 1968. Joint Automatic Control Conference, University of Michigan, Amer. Soc. Mech. Engrs. 1968, pp. 40—86.
4 PHILLIPS, R. S., Perturbation theory for semi-groups of linear operators, Trans. Amer. Math. Soc., 74 (1953), pp. 199—221.
5 GOPALSAMY, K., and BHARUCHA-REID, A. T., On a class of partial differential equations driven by stochastic point processes, J. App. Prob. (in press).
6 GOPALSAMY, K., On the conditional density of a noise driven evolution, ZAMM 56, 239—242 (1976)
7 GOPALSAMY, K., Harmonic analysis of plates on random elastic foundations, Proc. 4th Canadian Cong. Appl. Mech., May 1973, pp. 359—360.
8 PARZEN, E., Stochastic processes, Holden-Day, 1964, pp. 78—86.

Eingereicht: 11. 11. 1974

Anschrift: Prof. K. GOPALSAMY, School of Mathematical Sciences, Flinders University, Bedford Park, S.A. 5042, Australia

ZAMM 56, 461–468 (1976)

D. J. Temperley

Alternative Approaches to the Asymptotic Solution of $\varepsilon\nabla^2 u = \partial u/\partial y$, $0 < \varepsilon \ll 1$, over a Rectangle

Durch die asymptotische Lösung der Ausgangsgleichung mit den entsprechenden Randwerten für ein Rechteck $0 \leqq x \leqq l$, $0 \leqq y \leqq 1$ kann die Lösung für parabolische Grenzschichten an den Seitenwänden $x = 0, l$ für alle Ordnungen ohne direkten Bezug auf die Grenzschichten in den Ecken des Rechtecks erhalten werden. Drei mögliche Methoden zur Herleitung einer derartigen Lösung werden umrissen und der Näherung von Cook, Ludford und Walker [1] gegenübergestellt. Es wird gezeigt, daß die Verfahren von Temperley und Todd [2] zur Lösung des Ausgangsproblems für alle Bereiche des Rechtecks geeignet sind. Einige Bemerkungen zu dem entsprechenden MHD-Kanalproblem beschließen die Diskussion.

In the asymptotic solution of the title equation with prescribed boundary data over a rectangle $0 \leqq x \leqq l$, $0 \leqq y \leqq 1$, the solution in the parabolic boundary-layers on the side walls $x = 0, l$ may be obtained to all orders without explicit reference to the boundary-layers at the corners of the rectangle. Three possible methods of deriving such a solution are outlined and contrasted with the approach of Cook, Ludford and Walker [1]. The techniques described in Temperley and Todd [2] are shown to be applicable to the solution of the title problem over all parts of the rectangle. Some comments on the analogous MHD duct problem conclude the discussion.

С помощью асимптотического решения исходного уравнения с соответствующими краевыми значениями для прямоугольника $0 \leqq x \leqq l$, $0 \leqq x \leqq 1$ может быть получено решение для параболических пограничных слоёв на боковых стенках $x = 0, l$ всех порядков, причём пограничные слои в углах прямоугольника не имеют непосредственного значения. Описываются три возможных метода для получения такого решения, которое сравнивается с приближённым решением, полученным в работе Кук, Ладфорд и Вокер [1]. Показывается также, что метод, предлагаемый в работе Темперлей и Тодд [2] представляется подходящим для решения исходной проблемы для любых областей прямоугольника. В заключение рассматриваются некоторые замечания к соответствующей проблеме MHD течения в канале.

1. Introduction

In a recent paper Cook, Ludford and Walker [1] considered the asymptotic solution of the equation

$$(\varepsilon\nabla^2 - \partial/\partial y)\, u = 0\,, \qquad 0 < \varepsilon \ll 1\,, \tag{1}$$

over the rectangle $0 \leqq x \leqq l$, $0 \leqq y \leqq 1$, under the boundary conditions

$$u(x, 0) = 1, \quad u(x, 1) = -1 \qquad \text{for } 0 < x < l\,, \tag{2}$$

$$u(0, y) = u(l, y) = 0 \qquad \text{for } 0 < y < 1\,. \tag{3}$$

An analogy was drawn with the well-documented problem of magnetohydrodynamic flow down a rectangular duct of constant cross-section $0 \leqq |x| \leqq l$, $0 \leqq |y| \leqq 1$ (see, for example, Temperley and Todd [2]) governed by the equations

$$(M^{-1}\nabla^2 - \partial/\partial y)\, u_1 = -M^{-1} \tag{4}$$

and

$$(M^{-1}\nabla^2 + \partial/\partial y)\, u_2 = -M^{-1}\,, \tag{5}$$

where u_1, u_2 are the respective difference and sum of the non-dimensionalized induced velocity and magnetic fields measured down the axis of the duct and $M \gg 1$ is the Hartmann number (c. f. ε^{-1} in equation (1)).

Cook, Ludford and Walker argued that the bottom corners of the rectangle λ, far from being passive, actually generate the parabolic boundary-layers on the side walls $x = 0, l$ and claimed that the asymptotic solution in such a layer could only logically be obtained to all orders by expanding out the complete asymptotic approximation to the exact solution in the corresponding bottom corner; the top corners, though passive, lead to unusual elliptic quarter-plane problems. They strongly criticised the claim of Temperley and Todd [2] that the latter's method of matched asymptotic expansions (as applied to the MHD problem) could be pursued to all orders in regions distant $\gg \varepsilon$ from the corners without reference to the precise character of the approximations in the corners, and contended that only a remote possibility exists of choosing the correct approximation at each stage after the first without looking at the corners.

In this paper we affirm and shall prove that the title problem may be solved to all orders at every point outwith the boundary-layers at the top corners of the rectangle *without explicit reference to the situation in those corners*, which may be considered later as a separate exercise (see sections 4, 5). We deny that first to solve the problem in a region distant $O(\varepsilon)$ from either of the bottom corners and then to expand that solution in the corresponding side layer (the region distant $O(\varepsilon^{1/2})$ from the side wall but $\gg \varepsilon$ from the top wall) in terms of appropriate boundary-layer coordinates is the only logical way of obtaining the asymptotic expansion for u in the side layer.

We shall outline three alternative methods of solution, two of which involve the explicit use of a Minimum Singularity Principle; such a principle is 'built-in' to the third method which automatically yields a unique (and correct) solution to all orders in every part of the rectangle. The reader may compare and contrast each method with that adopted in [1].

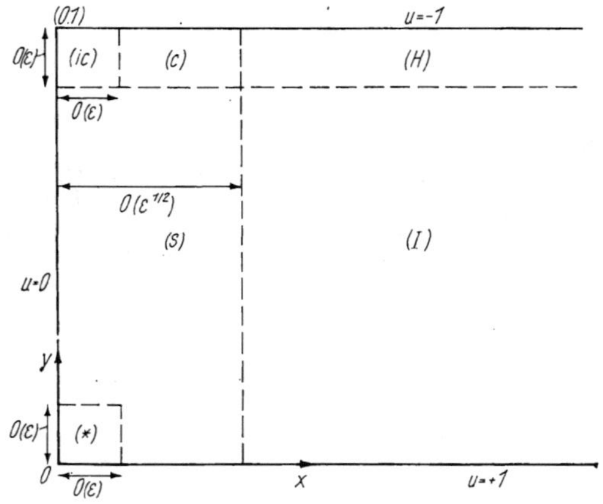

Fig. 1. Boundary-layers in the left-hand half of the rectangle for $0 < \varepsilon \ll 1$ (not to scale)

2. Sub-division of the rectangle

By symmetry, only the left-hand half of the rectangle need be considered (see Fig. 1). In addition to the interior (I) region four types of boundary-layer occur; these are denoted by (H), (s), (c) and (ic) to facilitate comparison or with the Hartmann, side, corner and inner-corner regions in the typical MHD duct cross-section considered by Temperley and Todd [2]. Region $(*)$ is not a proper boundary-layer, in that the solution for u in $(*)$ contains no terms peculiar to that region, but features in the investigation (see section 3).

In the (I) and (H) regions the solution, correct to asymptotically exponentially small terms in ε, is (see Cook, Ludford and Walker ([1], p. 118))

$$u \sim 1 - 2e^{-Y}, \text{ where } Y = (1 - y)/\varepsilon . \tag{6}$$

This solution, of course, contravenes boundary condition (3) and appropriate forms for u in the (s), (c) and (ic) regions must now be found.

3. The parabolic side layer

The appropriate horizontal boundary-layer coordinate being $X = x/\varepsilon^{1/2}$, we expand u asymptotically in the form

$$u(X, y, \varepsilon) \sim \sum_{n=0}^{\infty} u_n(X, y)\, \varepsilon^n . \tag{7}$$

Substituting into equation (1) with conditions (2), (3) and matching with result (6) yields the recurrence relation

$$(\partial^2/\partial X^2 - \partial/\partial y)\, u_n = -\, \partial^2 u_{n-1}/\partial y^2 \tag{8}$$

with

$$u_n(0, y) = 0 \ (n \geqq 0) \quad \text{for } 0 < y < 1 , \tag{9}$$

$$u_0(X, 0) = 1, \quad u_n(X, 0) = 0 \ (n \geqq 1) \quad \text{for } 0 < X < \infty \tag{10}$$

and

$$u_0 \to 1, \ u_n \to 0 \ (n \geqq 1) \quad \text{as } X \to \infty, \text{ for } 0 < y < 1 . \tag{11}$$

Cook, Ludford and Walker [1] quoted the error function solution to the problem for u_0, explaining how the various y-derivatives of the error function (eigensolutions of the boundary-value problem) are excluded by a Minimum Singularity argument, and produced a particular integral u_{1P} and complementary function u_{1C} to the problem for u_1 which were then matched by the same sort of argument. They contended that one would need to be remarkably perceptive in order to realise that both u_{1P} and u_{1C} behave like $\delta'(x)$ as $y \to 0$ (and presumably even more so in dealing with similar situations arising in the later problems for u_n, $n \geqq 2$) and hence suggested abandoning this line of approach. An alternative method which involves solving the full problem for $u = u^*(X^*, y^*)$ in the region $(*)$, where $X^* = x/\varepsilon$ and $y^* = y/\varepsilon$, and expanding u^* inside the side layer in terms of $X = X^*\varepsilon^{1/2}$, $y = y^*\varepsilon$ was put forward as the only rational way of obtaining the series form (7) for u in the (s) region.

We will show that, aside from the fact that region (*) is not a proper boundary-layer, it is not necessary to first solve for u^* before treating the side layer problem; three alternative methods will be outlined (in ascending order of simplicity), all of which avoid explicit investigation of region (*) and yet yield the asymptotic solution for u in (s) to all orders.

(i) LAPLACE transform method

Introducing

$$\tilde{u}_0(X, s) = \mathscr{L}\big(u_0(X, y)\big) = \int_0^\infty u_0 e^{-sy}\, dy\,,$$

multiplying equation (8) (with $n = 0$) by e^{-sy} and integrating from $y = 0$ to $y = \infty$ (one may treat the range of y as infinite and then 'patch' onto $y = 1$ via correction terms peculiar to the (c) and (ic) layers) we obtain

$$d^2\tilde{u}_0/dX^2 - s\tilde{u}_0 = -1\,. \tag{12}$$

The solution satisfying

$$\tilde{u}_0(0, s) = \mathscr{L}\big(u_0(0, y)\big) = 0 \tag{13}$$

is

$$\tilde{u}_0 = \big(1 - \exp\,(-Xs^{1/2})\big)/s\,. \tag{14}$$

Whilst it may be readily be shown that assumption (13) is valid, similar results do not apply in cases where $n \geqq 1$ (see later investigation of the problem for u_1, for example).

Inverting result (14) yields (see ERDELYI et al. [3], p. 176)

$$u_0 = \operatorname{erf}\,(X/2y^{1/2})\,. \tag{15}$$

The solution contains no multiple of any y-derivative of the error function; an automatic Minimum Singularity Principle is 'built-in' via assumption (13) which precludes singularities in u_0 of type $\delta^{(m)}(X)$, $m \geqq 0$ and integral, as $y \to 0$. However, the solution for u_1 is not so reasonably behaved near the origin and a more explicit Minimum Singularity Principle must be applied.

Setting $n = 1$ in lines (8)—(11), introducing

$$\tilde{u}_1(X, s) = \int_0^\infty u_1(X, y)\, e^{-sy}\, dy$$

and proceeding as before yields

$$d^2\tilde{u}_1/dX^2 - s\tilde{u}_1 = -s^2\tilde{u}_0 = s\exp\,(-Xs^{1/2}) \tag{16}$$

and hence

$$\tilde{u}_1 = -\tfrac{1}{2} s^{1/2} X e^{-Xs^{1/2}} + A(s)\, e^{-Xs^{1/2}}\,. \tag{17}$$

If one were now to assume (as in (13)) that

$$A(s) = \tilde{u}_1(0, s) = \mathscr{L}\big(u_1(0, y)\big) = 0\,, \tag{18}$$

the resulting (incorrect) solution would be (see [3], p. 246)

$$u_1 = \mathscr{L}^{-1}\,(-\tfrac{1}{2} s^{1/2} X e^{-Xs^{1/2}}) = X(2y - X^2)\exp\,(-X^2/4y)/8\pi^{1/2}y^{5/2} = U_1(X, y)\,. \tag{19}$$

Instead of proceeding more cautiously, one must first expand $A(s)$ in the form $A(s) = A_0 + A_1 s + A_2 s^2 + \ldots$, the A_i's being constants, so that

$$u_1 = U_1(X, y) + \sum_{i=0}^\infty A_i\, \mathscr{L}^{-1}\,(s^i e^{-Xs^{1/2}})\,. \tag{20}$$

Near the origin the second and all subsequent terms in the infinite series may readily be shown to exhibit behaviour more singular than $U_1(X, y)$ and so, applying a Minimum Singularity Principle, we exclude them by setting $A_i = 0$, all $i \geqq 1$. Thus (see [3], p. 245),

$$u_1 = U_1(X, y) + A_0 X\exp\,(-X^2/4y)/2\pi^{1/2}y^{3/2}\,. \tag{21}$$

As $y \to 0$ in the latter result, $U_1(X, y)$ behaves like $2\delta'(X)$ while the second term behaves like $-2A_0\delta'(X)$ (see Appendix A); we therefore choose $A_0 = 1$, in order that the two singularities may annihilate one another at the origin, and the resulting solution (21) now matches that obtained by COOK, LUDFORD and WALKER [1]. The comments in [1] about the need to be somewhat clever in order to 'neutralise' the singularities of the two parts of expression (21) presumably pertain to the analysis set out in the Appendix A. Removal of the $\delta'(X)$-type singularity as $y \to 0$ by the above procedure introduces, however, a singularity $2\delta(y)$ into u_1 as $X \to 0$, a singularity of a type not present in the provisional solution $U_1(X, y)$ (see Appendix B).

The asymptotic expansion for u may be obtained to all orders in the (s) layer using the above method. At the $O(\varepsilon^n)$ level an assumption of the form (18) leads to a solution $U_n(X, y)$ which behaves like $\delta^{(2n-1)}(X)$ as $y \to 0$; the addition of the appropriate eigen-solution annihilates this singularity but introduces into u_n a singularity of the type $\delta^{(n-1)}(y)$ as $X \to 0$, i.e. the Minimum Singularity principle operates in such a way as to delete the high order singularity occurring as $y \to 0$, replacing it with a lower order one which arises when $X \to 0$.

(ii) FOURIER sine transform method

Returning to the problem for u_0 and introducing

$$\bar{u}_0(\lambda, y) = \int_0^\infty u_0(X, y) \sin(\lambda X)\, dX ,$$

we multiply the governing equation by $\sin(\lambda X)$ and then integrate from $X = 0$ to $X = \infty$ to obtain

$$d\bar{u}_0/dy + \lambda^2 \bar{u}_0 = 0 . \tag{22}$$

The appropriate solution which satisfies $\bar{u}_0(\lambda, 0) = 1/\lambda$ is

$$\bar{u}_0 = \exp(-\lambda^2 y)/\lambda \tag{23}$$

and hence (see [3], p. 96)

$$u_0(X, y) = 2/\pi \int_0^\infty \bar{u}_0 \sin(\lambda X)\, d\lambda = \operatorname{erf}(X/2y^{1/2}) . \tag{24}$$

We have here used the result that the boundary value of the FOURIER sine transform of u_n equals the transform of the boundary value, which holds for all $n \geqq 0$.

Introducing $\bar{u}_1(\lambda, y) = \int_0^\infty u_1(X, y) \sin(\lambda X)\, dX$ and proceeding as above we find that

$$d\bar{u}_1/dy + \lambda^2 \bar{u}_1 = d^2\bar{u}_0/dy^2 = \lambda^3 \exp(-\lambda^2 y) . \tag{25}$$

The solution satisfying $\bar{u}_1(\lambda, 0) = 0$ is

$$\bar{u}_1 = \lambda^3 y \exp(-\lambda^2 y) \tag{26}$$

and hence (see GRADSHTEYN and RYZHIK [4], p. 495)

$$u_1 = X(6y - X^2) \exp(-X^2/4y)/8\pi^{1/2} y^{5/2} . \tag{27}$$

One may continue indefinitely in this manner, deriving the asymptotic expansion for u in the (s) layer to all orders. For example,

$$\bar{u}_2(\lambda, y) = \int_0^\infty u_2(X, y) \sin(\lambda X)\, dX = \tfrac{1}{2} \lambda^5 y (\lambda^2 y - 4) \exp(-\lambda^2 y) \tag{28}$$

and hence (see [4], p. 496, result 10 and p. 1033, para 8.950)

$$\begin{aligned} u_2 &= -\{H_7(X/2y^{1/2}) + 16 H_5(X/2y^{1/2})\} \exp(-X^2/4y)/2^8 \pi^{1/2} y^2 \\ &= -X(X^6 - 26X^4 y + 100 X^2 y^2 + 120 y^3) \exp(-X^2/4y)/2^8 \pi^{1/2} y^{11/2} . \end{aligned} \tag{29}$$

Similarly

$$\bar{u}_3(\lambda, y) = \tfrac{1}{6} \lambda^7 y (\lambda^4 y^2 - 12\lambda^2 y + 30) \exp(-\lambda^2 y) , \tag{30}$$

and hence

$$u_3 = -\frac{X}{3.2^{12}\pi^{1/2} y^{17/2}} (X^{10} - 62X^8 y + 984 X^6 y^2 - 3024 X^4 y^3 - 5040 X^2 y^4 - 10070 y^5) \exp(-X^2/4y) . \tag{31}$$

Method (ii) produces a unique, correct solution to all orders *automatically*, without recourse to the type of complicated analysis deemed essential by COOK, LUDFORD and WALKER; a Principle of Minimum Singularity is built-in to the method due to the properties of the FOURIER sine transform with respect to X, which cuts out the singularity of the type $\delta^{(2n-1)}(X)$ in u_n as $y \to 0$ whilst leaving behind one proportional to $\delta^{(n-1)}(y)$ as $X \to 0$. It is certainly not *necessary* to introduce the solution u^* from which the (s) layer solution was constructed in [1]. Nor is any explicit reference required to the (c) and/or (ic) layers; a separate analysis (see sections 4,5) yields the correction terms peculiar to the latter regions.

(iii) inductive method

Professor L. TODD[1]) has obtained the asymptotic solution for u to all orders in the side layer as follows.

The governing equation for u_1 being

$$(\partial^2/\partial X^2 - \partial/\partial y)\, u_1 = -\partial^2 u_0/\partial y^2 = -\partial^2/\partial y^2 \left(\operatorname{erf}(X/2y^{1/2})\right) , \tag{32}$$

it is obvious that a particular integral of the form

$$u_{1p} = B_1 y \partial^2/\partial y^2 \left(\operatorname{erf}(X/2y^{1/2})\right)$$

[1]) Professor L. TODD, Department of Mathematics, Laurentian University, Sudbury (Canada)

is available and one may readily deduce that $B_1 = 1$. All y-derivatives of erf $(X/2y^{1/2})$ being more singular at the origin than u_{1p} we conclude that the complementary function $u_{1c} \equiv 0$ and hence

$$u_1 = y\partial^2/\partial y^2 \left(\operatorname{erf}(X/2y^{1/2})\right). \tag{33}$$

Further, the equation

$$(\partial^2/\partial X^2 - \partial/\partial y)\, u_2 = -\, \partial^2 u_1/\partial y^2 = -\,(2\partial^3/\partial y^3 + y\partial^4/\partial y^4)\operatorname{erf}(X/2y^{1/2}) \tag{34}$$

has a particular integral of the form

$$u_{2p} = (B_2 y\partial^3/\partial y^3 + B_3 y^2\partial^4/\partial y^4)\operatorname{erf}(X/2y^{1/2})$$

and one may readily deduce that $B_2 = 2,\ B_3 = 1/2$. Thus

$$u_{2p} = -\,\frac{X}{2^8\pi^{1/2}y^{11/2}}\,(X^6 - 26X^4y + 100X^2y^2 + 120y^3)\exp(-\,X^2/4y)\,, \tag{35}$$

which has no singularities of the type $\delta(X)$, $\delta'(X)$ or worse as $y \to 0$. It should be noted, however, that u_{2p} behaves like $\delta'(y)$ as $X \to 0$. By a Minimum Singularity argument, no y-derivative of erf $(X/2y^{1/2})$ can be included in the complementary function which hence vanishes identically. Thus $u_2 \equiv u_{2p}$ which matches result (29).

Proceeding further, it is readily shown that

$$u_3 \equiv u_{3p} = (5y\partial^4/\partial y^4 + 2y^2\partial^5/\partial y^5 + \tfrac{1}{6}\,y^3\partial^6/\partial y^6)\operatorname{erf}(X/2y^{1/2}) \tag{36}$$

and

$$u_4 \equiv u_{4p} = (14y\partial^5/\partial y^5 + 7y^2\partial^6/\partial y^6 + y^3\partial^7/\partial y^7 + \tfrac{1}{24}\,y^4\partial^8/\partial y^8)\operatorname{erf}(X/2y^{1/2})\,, \tag{37}$$

the complementary function being identically zero in each case by the Minimum Singularity Principle.

We postulate that $u_{nc} \equiv 0$ (all $n \geqq 1$) and that the u_n all take the form

$$u_n \equiv u_{np} = (B_m y\partial^{n+1}/\partial y^{n+1} + B_{m+1}y^2\partial^{n+2}/\partial y^{n+2} + \cdots + B_{m+n-1}y^n\partial^{2n}/\partial y^{2n})\operatorname{erf}(X/2y^{1/2})\,, \tag{38}$$

where $m = (n^2 - n + 2)/2$.

Thus the asymptotic expansion for u in (s) may be obtained to all orders without explicit reference to regions (*), (c) or (ic). Method (iii) is the most elegant yet devised and again provides a complete asymptotic solution throughout the (s) region.

Since the (s) layer solution clearly contravenes boundary condition (2) at the upper boundary of the (c) and (ic) layers, we must now consider the correct form of asymptotic expansion appropriate to those regions.

4. The corner layer

In this region we add to the solution applicable outwith the (c), (ic) layers a correction term

$$u_c(X, Y, \varepsilon) = \sum_{n=0}^{\infty} u_c^{(n)}(X, Y)\,\varepsilon^n\,, \tag{39}$$

which decays exponentially as X and/or $Y \to \infty$.

Outside the (c) and (ic) layers, u takes the form $u = \hat{u} = \sum_{n=0}^{\infty} \hat{u}_n\varepsilon^n$ with (see results (6), (15))

$$\hat{u}_0 = -\,2e^{-Y} + \operatorname{erf}(X/2y^{1/2})\,. \tag{40}$$

Substituting in equation (1) and condition (2) we have, at the $O(\varepsilon^\circ)$ level,

$$(\partial^2/\partial Y^2 + \partial/\partial Y)\, u_c^{(0)} = 0 \tag{41}$$

with

$$u_c^{(0)}(X, 0) = -\,1 - (\hat{u}_0)_{y=1} = 1 - \operatorname{erf}(X/2)\,, \tag{42}$$

and hence

$$u_c^{(0)} = \left(1 - \operatorname{erf}(X/2)\right)e^{-Y}\,. \tag{43}$$

The problem for $u_c^{(1)}$ is

$$(\partial^2/\partial Y^2 + \partial/\partial Y)\, u_c^{(1)} = -\,\partial^2 u_c^{(0)}/\partial X^2 = X\exp(-\,Y - X^2/4)/2\pi^{1/2} \tag{44}$$

with (see result (27))

$$u_c^{(1)}(X, 0) = -\,\hat{u}_2(X, 1) = X(X^2 - 6)\exp(-\,X^2/4)/8\pi^{1/2}\,. \tag{45}$$

Thus $u_c^{(1)} = X(X^2 - 4Y - 6)\exp(-\,Y - X^2/4)/8\pi^{1/2}\,.$ (46)

Similarly

$$(\partial^2/\partial Y^2 + \partial/\partial Y)\, u_c^{(2)} = -\,\partial^2 u_c^{(1)}/\partial X^2$$

with $u_c^{(2)}(X, 0) = -\hat{u}_2(X, 1)$, and hence (see results (46), (29))

$$u_c^{(2)} = \frac{X}{2^8\pi^{1/2}} \{X^6 + 2X^4(4Y - 13) + 4X^2(25 - 48Y - 4Y^2) + 24(5 + 28Y + 4Y^2)\} e^{-(Y + X^2/4)} . \tag{47}$$

One can proceed in this manner indefinitely, obtaining (c) layer correction terms to any order which may then be 'patched' onto the $\hat{u}_n$ components from section 3.

5. The inner-corner layer

The solution $u = \hat{u} + u_c$ contravenes boundary condition (3) at the left-hand boundary of the (ic) layer. We therefore add on a further correction term

$$u_{ic}(X^*, Y, \varepsilon) = \sum_{n=0}^{\infty} u_{ic}^{(n)}(X^*, Y)\, \varepsilon^n \tag{48}$$

which decays exponentially as X^* and/or $Y \to \infty$. From equation (1), conditions (2/3) and results (40), (43) it follows that

$$(\nabla^2 + \partial/\partial Y)\, u_{ic}^{(0)} = 0 \tag{49}$$

with

$$u_{ic}^{(0)}(X^*, 0) = 0, \quad u_{ic}^{(0)}(0, Y) = -(\hat{u}_0)_{x=0} - u_c^{(0)}(0, Y) = e^{-Y} . \tag{50}$$

Introducing

$$\bar{u}_{ic}(k, Y) = \int_0^{\infty} u_{ic}^{(0)}(X^*, Y) \sin(kX^*)\, dX^* ,$$

multiplying equation (49) by $\sin(kX^*)$ and integrating from $X^* = 0$ to $X^* = \infty$ yields

$$(d^2/dY^2 + d/dY - k^2)\, \bar{u}_{ic} = -k\, e^{-Y} . \tag{51}$$

The solution satisfying $\bar{u}_{ic} \to 0$ as $Y \to \infty$ and $\bar{u}_{ic}(k, 0) = 0$ is

$$\bar{u}_{ic} = \{\exp(-Y) - \exp(-Y(1 + (1 + 4k^2)^{1/2})/2)\}/k . \tag{52}$$

Integration with respect to y of result (26), p. 16 of [3] enables the inverse transform to be written in the form

$$u_{ic}^{(0)} = e^{-Y} - Y e^{-Y/2}/\pi \int_0^{X^*} K_1\{(\theta^2 + Y^2)^{1/2}/2\}/(\theta^2 + Y^2)^{1/2}\, d\theta , \tag{53}$$

which matches result (7) in [1].

It may readily be shown that $u_c^{(n)}(0, Y) = 0$ (all $n \geqq 1$) and hence, since $(\hat{u}_n)_{x=0}$ is zero for all $n \geqq 1$ it follows that $u_{ic}^{(n)}(0, Y) = 0$, all $n \geqq 1$. Thus

$$u_{ic}^{(n)} \equiv 0, \text{ all } n \geqq 1 , \tag{54}$$

only the additional $u_{ic}^{(0)}$ correction term being required in order to 'patch' the $\{\hat{u} + u_c\}$ solution onto the top section of the left-hand wall.

Our transform techniques thus produce the solution to the title problem without any detailed investigation of region (*).

6. The MHD duct problem

The techniques used by COOK, LUDFORD and WALKER to solve the title problem are not, broadly speaking, applicable to the corresponding MHD duct problem (see TEMPERLEY and TODD [2]). In order to derive the solution in the (s) layer one would first have to solve the inner-corner problem to all orders and this can only be achieved in cases where the duct walls parallel to the externally applied field are non-conducting; for any other configuration the method is inappropriate. The only scheme applicable to all configurations is the matched asymptotic expansion procedure adopted by TEMPERLEY and TODD [2], whereby the (ic) solution can only be constructed to the same level as that in the (s) and (c) layers, which must be examined in conjunction (but can be examined independent of the (ic) region).

7. Final Comments

COOK, LUDFORD and WALKER asserted that it was most unlikely that one could choose the correct approximation at each stage after the first in the MHD side layer without looking at the corners and sought to justify this assertion by reference to the (simpler) title problem. However, we have shown that their own method is far from being the only way of solving the title problem and it is not applicable to the general MHD duct problem. The expansion procedure of TEMPERLEY and TODD (method (ii)), on the other hand, yields the correct asymptotic solution to the title problem to all orders at every point in the rectangle without recourse to the introduction of explicit Minimum Singularity arguments. In this respect, the use of FOURIER sine transforms with respect to the horizontal boundary-layer coordinate scores heavily over method (i), since the boundary-value of the LAPLACE transform cannot be automatically matched with the transform of the corresponding boundary-value of the dependent variable without leaving in the highest order singularity at all levels after the first.

The key question, one we are still pursuing, is whether the TEMPERLEY-TODD solutions in the MHD (s) and (c) layers for various configurations are unique and correct; this is undoubtedly the case at any given level *provided they do not exhibit bad behaviour in the* (ic) *region.* At a certain stage, non-uniqueness plus 'bad inner-corner behaviour' difficulties may arise in the side layer solution and it would clearly be better, if feasible, to first obtain the asymptotic solution in the (ic) region and then expand that solution out as suggested by COOK, LUDFORD and WALKER.

We concede, then, that the MHD(s) layer solutions may require to be matched with the (ic) solutions in an unusual way at a certain stage and our scheme would then require modification. It may be wrong, for instance, to assume transcendental decay of components outside their regions of influence (see [2], section 4) and to treat the (ic) solution as being composed merely of the sum of the solution outside the (ic) regions plus exponentially decaying terms confined to the inner-corners.

We are at present working on a method of obtaining the solution for $M \gg 1$ in closed form for a duct having non-conducting walls parallel to the imposed magnetic field; this would hold in all boundary-layer regions and improve on any algebraic series.

Appendix A

Form taken by expression (21) *as* $y \to 0$.
For any ordinary function $F(X)$ (see LIGHTHILL [5], p. 11)

$$\int_{-\infty}^{\infty} \delta'(X)\, F(X)\, \mathrm{d}X = - \int_{-\infty}^{\infty} \delta(X)\, F'(X)\, \mathrm{d}X = - F'(0)\,. \tag{55}$$

Combining each part of expression (21) in turn with such a function $F(x)$ we have

$$\text{(a)} \qquad \int_{-\infty}^{\infty} U_1(X, y)\, F(X)\, \mathrm{d}X = \int_{-\infty}^{\infty} \frac{t(1-2t^2)}{(\pi y)^{1/2}}\, \mathrm{e}^{-t^2} F(2ty^{1/2})\, \mathrm{d}t\,, \qquad t = X/2y^{1/2}\,,$$

$$= -1/\pi^{1/2} \int_{-\infty}^{\infty} (1 + 2t^2)\, \mathrm{e}^{-t^2} F'(2ty^{1/2})\, \mathrm{d}t$$

$$\to -F'(0)/\pi^{1/2} \int_{-\infty}^{\infty} (1+2t^2)\, \mathrm{e}^{-t^2}\, \mathrm{d}t = -2F'(0) \quad \text{as} \quad y \to 0\,. \tag{56}$$

$$\text{(b)} \qquad \int_{-\infty}^{\infty} A_0 X \exp(-X^2/4y)\, F(X)/(2\pi^{1/2} y^{3/2})\, \mathrm{d}X = \frac{2A_0}{(\pi y)^{1/2}} \int_{-\infty}^{\infty} t\, \mathrm{e}^{-t^2} F(2ty^{1/2})\, \mathrm{d}t$$

$$= \frac{2A_0}{\pi^{1/2}} \int_{-\infty}^{\infty} \mathrm{e}^{-t^2} F'(2ty^{1/2})\, \mathrm{d}t\,,$$

$$\to 2A_0 F'(0) \quad \text{as} \quad y \to 0\,. \tag{57}$$

Thus the two terms in expression (21) behave like $2\delta'(X)$, $-2A_0\,\delta'(X)$ respectively as $y \to 0$.

Appendix B

Form taken by expression (21) *as* $X \to 0$
Combining each part of expression (21) with any ordinary function $G(y)$, u_1 being defined for $y < 0$ by $u_1(X, -y) = +u_1(X, y)$, we have

$$\text{(i)} \qquad I_1 = \int_{-\infty}^{\infty} G(y)\, U_1(X, y)\, \mathrm{d}y = 2 \int_{0}^{\infty} G_e(y)\, X(2y - X^2) \exp(-X^2/4y)/8\pi^{1/2} y^{5/2}\, \mathrm{d}y\,,$$

where $G_e(y) \equiv \frac{1}{2}\{G(y) + G(-y)\}$, the even part of $G(y)$.

$$\text{i.e.} \qquad I_1 = 2/\pi^{1/2} \int_{0}^{\infty} G_e(X^2/4t^2)\, \mathrm{e}^{-t^2} (1 - 2t^2)\, \mathrm{d}t\,, \qquad t = X/2y^{1/2}\,.$$

$$\to 2G_e(0)/\pi^{1/2} \int_{0}^{\infty} \mathrm{e}^{-t^2} (1 - 2t^2)\, \mathrm{d}t = 0\,, \qquad \text{as } X \to 0\,. \tag{58}$$

Thus $U_1(X, y)$ has no singularities of the type $\delta(y)$, $\delta'(y)$ or worse as $X \to 0$.

$$\text{(ii)} \qquad I_2 = \int_{-\infty}^{\infty} G(y)\, A_0 X \exp(-X^2/4|y|)/2\pi^{1/2} |y|^{3/2}\, \mathrm{d}y = 4A_0/\pi^{1/2} \int_{0}^{\infty} G_e(X^2/4t^2)\, \mathrm{e}^{-t^2}\, \mathrm{d}t$$

$$\to 4A_0\, G_e(0)/\pi^{1/2} \int_{0}^{\infty} \mathrm{e}^{-t^2}\, \mathrm{d}t = 2A_0\, G(0) \quad \text{as } X \to 0\,. \tag{59}$$

The addition of the relevant eigen-function (corresponding to $A_0 = 1$) thus introduces a singularity $2\delta(y)$ into u_1 as $X \to 0$ (see result (55)).

If, on the other hand, we were to extend the definition of u_1 so that u_1 was an odd function of y, no singularities of the type $\delta(y)$, $\delta'(y)$ or worse would occur in u_1 as $X \to 0$. Defining $u_1(X, y) \equiv 0$ for $y < 0$ leaves result (58) unchanged but the eigen-function now tends to $\delta(y)$ as $X \to 0$.

References

1 COOK, L. P., LUDFORD, G. S. S., WALKER, J. S., Corner regions in the asymptotic solution of $\varepsilon\nabla^2 u = \partial u/\partial y$ with reference to MHD duct flow, Proc. Cambridge Philos. Soc. **72**, 117–121 (1972).
2 TEMPERLEY, D. J., TODD, L., The effects of wall conductivity in magnetohydrodynamic duct flow at high HARTMANN number, Proc. Cambridge Philos. Soc. **69**, 337–351 (1971).
3 ERDÉLYI, A. et al, Tables of Integral Transforms. Vol. 1, McGraw-Hill, New York 1954.
4 GRADSHTEYN, I. S., RYZHIK, I. M., Tables of Integrals, Series and Products, Academic Press, New York 1965.
5 LIGHTHILL, M. J., Introduction to FOURIER Analysis and Generalised Functions, Cambridge University Press, Cambridge 1960.

Eingereicht: 27. 6. 1975

Anschrift: Dr. D. J. TEMPERLEY, Department of Applied Mathematics, University of Edinburgh, Edinburgh, Scotland.

ZAMM 56, 469–478 (1976)

L. DEBNATH

On Wind-Driven Currents in an Ocean with Bottom Friction

Es wird die Langperiodenreaktion eines unbegrenzten β-ebenen Ozeans mit Bodenreibung auf Anfangsstörungen durch Winddruck auf die Oberfläche untersucht. Zur Beschreibung der freien und erzwungenen barotropen Bewegungen, die von stationären oder wandernden atmosphärischen Winden erzeugt werden, wird ein Verfahren entwickelt. Die Approximation der stationären Phase wird zur Berechnung der asymptotischen Lösung, die den Endcharakter der ozeanischen Bewegung beschreibt, genutzt.

This paper presents an investigation of the long period response of an unbounded β-plane ocean with bottom friction to initial disturbances of wind stress applied at the surface. An analysis is carried out for determining the free and forced barotropic motions generated by stationary or traveling atmospheric winds. The stationary phase approximation is employed to calculate the asymptotic solution which describes the ultimate feature of the oceanic motion. It is shown that the concentrated initial disturbance ultimately produces two systems of waves-transverse and longitudinal wave systems which are radically modified by the bottom friction. In contrast to the solution in an ocean model without bottom friction, the amplitude of the transverse and the longitudinal waves decay exponentially as time $t \to \infty$. The barotropic motion produced by several physically realistic atmospheric wind distributions is also investigated. The main object of this preliminary study is to consider the simplest possible ocean model with linearized bottom friction for an understanding of its influence on the wind-induced ocean currents so that this study can be used as a guide for the investigation of the non-linear frictional model with or without viscous effect. The physical significance of the mathematical results of this analysis is discussed.

Исследуется реакция с большим периодом некоторого неограниченного β-плоского океана с трением на дне на начальные возмущения посредством давления ветра на поверхности. Для описания свободных и вынужденных баротропных движений, которые возникают из-за стационарных и блуждающих ветров, развивается некоторый метод. Аппроксиматция статционарной фазы используется для вычисления асимптотического решения, описывающего конечный характер движения океана.

1. Introduction

Based upon the pioneer works of ROSSBY [1—2] on the wind-driven ocean circulation, several authors including SVERDRUP [3], STOMMEL [4], CHARNEY [5], BOLIN [6], CAHN [7], VERONIS and STOMMEL [8], VERONIS [9—10] have made some remarkable progress towards the understanding of long time average, large and small time scale ocean currents generated by atmospheric winds, and the westward intensification of ocean currents observed in the Kuroshio and the Gulf Stream. STOMMEL has demonstrated that the variation of the Coriolis parameter with latitude — the so called β-effect is the main cause of the east-west asymmetry of the ocean circulation and has dominant influence on the dynamics of ocean. Especially, the formation of very intense currents (the Gulf Stream and others) in the western boundary of an ocean than in the east is essentially due to the β-effect.

Some of the above works have proved that the duration and the nature of the atmospheric disturbances have significant impact on the partition of input energy between the geostropic and non-geostrophic flows. Especially, it has been shown that when the time scale of the atmospheric disturbances is large compared to half pendulum day, the most of the input energy goes to the barotropic and baroclinic ROSSBY waves which constitute the so called geostrophic motions. On the other hand, when the wind stress is impulsive in nature or acts over a period less than a half pendulum day, then most of the energy is expended in the inertio-gravitational wave motions.

LONGUET-HIGGINS [11] has made an interesting study of the long period response of an unbounded non-uniformly rotating shallow ocean without bottom friction to initial distributions of stationary or traveling wind stress applied at the surface. Almost his entire analysis has been devoted to barotropic flows and then it has been applied to baroclinic motions with necessary modifications. He has also carried out an asymptotic analysis of the solution for large time, and shown that the initial disturbance eventually propagates in the form of two wave systems consisting of the transverse and longitudinal systems with amplitude which decays like t^{-1}. The response of the ocean to a moving storm traveling on a straight course has also been investigated and the resulting oceanic motions have been discussed in some detail.

Although all of the above works provide considerable amount of qualitative and quantitative information about the wind-driven ocean currents, but none of these studied have included the effects of viscosity, bottom friction and/or bottom topography on oceanic motions. Perhaps, there is no other justification for neglecting these effects than the fact that the analysis is quite complicated when these effects are included. It is, however, conceivable that these effects may be physically significant for real oceanic circulation. So the question of applicability of the models considered by the above authors to real oceans is still open for criticism. It may be anticipated that attempts to include all of the above effects together would lead to a tremendous analytical complexity and it would be a formidable task to solve the problem.

In the present paper, we propose to consider the wind-driven currents in a non-uniformly rotating shallow ocean with the dissipating effect due to the bottom friction. It is generally known that the effect of the bottom friction is reflected in the equations of motion through the frictional force $\varkappa f(\boldsymbol{u})$ where $\varkappa(> 0)$ is the coefficient of friction and $f(\boldsymbol{u})$ is a polynomial function of velocity. So the bottom friction introduces nonlinearity into the equations of motion and the resulting flow problem is in general difficult to solve. Physically, this dissipating frictional force appears to reduce the mass transport and in turn affects the oceanic motions produced by the wind stresses. Thus, the inclusion of this effect is of concrete physical significance and has practical applications to real oceans.

In this preliminary study, it is of interest to consider the simplest possible theoretical ocean model with linearized bottom friction for an understanding of its influence on the wind-induced ocean currents so that this study can be used as a guide for the investigation of the non-linear frictional model with or without viscous effects.

The main object of this paper is to make an investigation of the long period response of an unbounded β-plane ocean with bottom friction to initial distributions of wind stress applied at the surface. The analysis is carried out for determining the free and forced barotropic motions produced by stationary or traveling atmospheric winds. The stationary phase approximation is employed to describe the ultimate feature of the oceanic motion. The stationary phase method is employed to calculate the asymptotic solution which is found to describe the ultimate oceanic motion. It is shown that the concentrated initial disturbance ultimately produces two systems of waves, a transverse and a longitudinal wave system which are radically modified by the bottom friction. In contrast to the solution in an ocean model without bottom friction, the amplitude of the transverse as well as longitudinal wave system dies out exponentially as time $t \to \infty$. The barotropic motion produced by several physically realistic atmospheric wind distributions is also investigated. The physical significance of the mathematical results of this paper is discussed.

2. Mathematical Formulation

We consider an inviscid incompressible homogeneous ocean of uniform depth h and density ϱ under the influence of linearized bottom friction. It is of interest to investigate the dynamic response of the β-plane ocean model with turbulent momentum transfer neglected to a stationary or traveling wind stress applied at the free surface of the ocean.

We take the rectangular Cartesian coordinates x and y along the east and north directions respectively. We assume that the horizontal scale is sufficiently large so that the pressure can be described by the hydrostatic equation and the induced wind-driven motion is small so that the non-linear effects are negligible. We take into account that the Coriolis parameter $f = 2\Omega \sin \varphi$ has a positive gradient $\beta = \mathrm{d}f/\mathrm{d}y$ in the north direction, where Ω is the angular velocity of the earth's rotation and φ is the latitude of the place.

Based upon the well-known β-plane approximation, the horizontal velocity field (u, v) satisfies the linearized equations

$$\frac{\partial u}{\partial t} - fv + \varkappa u = -g\frac{\partial \zeta}{\partial x} + \frac{\tau^x}{\varrho h}, \tag{2.1}$$

$$\frac{\partial v}{\partial t} + fu + \varkappa v = -g\frac{\partial \zeta}{\partial y} + \frac{\tau^y}{\varrho h}, \quad (2.2), \qquad \frac{\partial \zeta}{\partial t} = -h\left(\frac{\partial u}{\partial x} + \frac{\partial v}{\partial y}\right), \tag{2.3}$$

where $\varkappa$ is the coefficient of friction involved in the frictional dissipative terms $\varkappa u$, $\varkappa v$; $\zeta(x, y, t)$ is the free surface elevation, g is the acceleration due to gravity, τ^x and τ^y are the components of wind stress at the surface. The elimination of ζ from equations (2.1)—(2.3) gives

$$\frac{\partial^2 u}{\partial t^2} - f\frac{\partial v}{\partial t} + \varkappa\frac{\partial u}{\partial t} = c^2\left(\frac{\partial^2 u}{\partial x^2} + \frac{\partial^2 v}{\partial x\, \partial y}\right) + \frac{\partial}{\partial t}\left(\frac{\tau^x}{\varrho h}\right), \tag{2.4}$$

$$\frac{\partial^2 v}{\partial t^2} + f\frac{\partial u}{\partial t} + \varkappa\frac{\partial v}{\partial t} = c^2\left(\frac{\partial^2 u}{\partial x\, \partial y} + \frac{\partial^2 v}{\partial y^2}\right) + \frac{\partial}{\partial t}\left(\frac{\tau^y}{\varrho h}\right), \tag{2.5}$$

where $c^2 = gh$. Eliminating u from (2.4) and (2.5), we obtain

$$\left[\left(\frac{\partial^2}{\partial t^2} + \varkappa\frac{\partial}{\partial t}\right)\nabla^2 - \frac{1}{c^2}\left\{\frac{\partial^4}{\partial t^4} + 2\varkappa\frac{\partial^3}{\partial t^3} + (f^2 + \varkappa^2)\frac{\partial^2}{\partial t^2}\right\} + \beta\frac{\partial^2}{\partial x\, \partial t}\right] v$$
$$= \frac{1}{\varrho h}\left[\frac{1}{c^2}\left(f\frac{\partial^2 \tau^x}{\partial t^2} - \frac{\partial^3 \tau^y}{\partial t^3} + \varkappa\frac{\partial^2 \tau^y}{\partial t^2}\right) + \frac{\partial^2}{\partial x\, \partial t}\left(\frac{\partial \tau^y}{\partial x} - \frac{\partial \tau^x}{\partial y}\right)\right]. \tag{2.6}$$

We examine the motions whose period is large compared to a pendulum day so that $\partial/\partial t \ll f$. We also assume that $(L^2 f^2)/c^2 \leqq O(1)$ so that $f\, \partial/\partial t \ll c^2\, \partial^2/\partial x^2$, where L is the horizontal length scale. Hence the highest derivatives with respect to time can be neglected on both sides of (2.6) and this means that the inertio-gravitational motions disappear from the analysis.

Integrating (2.6) once with respect to t and omitting the arbitrary function of y from v, it turns out that

$$\left[\nabla^2 - \frac{f^2 + \varkappa^2}{c^2}\right]\frac{\partial v}{\partial t} + \varkappa\nabla^2 v + \beta\frac{\partial v}{\partial x} = \frac{1}{\varrho h}\frac{\partial}{\partial x}\operatorname{curl}_z \tau, \tag{2.7}$$

where

$$\mathrm{curl}_z\,\tau \equiv \left(\frac{\partial \tau^y}{\partial x} - \frac{\partial \tau^x}{\partial y}\right). \tag{2.8}$$

It is noted that equation (2.7) describes the barotropic motions in a non-uniformly rotating shallow ocean with bottom friction. When the frictional dissipating effects are neglected, (2.7) reduces to an equation studied by LONGUET-HIGGINS [11]. When $f = 0$, (2.7) represents the propagation of well known non-dispersive long waves. If, however, f is constant, equation (2.7) describes dispersive but isotropic long wave motions. On the contrary, the variable Coriolis parameter, that is, the positive β-effect makes the wave motions not only dispersive but also anisotropic.

The main interest in this paper is to investigate the wave solutions of equation (2.7) without or with the various types of wind stress applied on the surface of the ocean and to examine the influence of the bottom friction on the wind-induced barotropic motions.

3. The Free Wave Solutions

In the absence of wind stress ($\tau^x = \tau^y = 0$), it is possible to find wave solutions of (2.7) proportional to $\exp[i(kx + ly - \omega t)]$ so that

$$v = V \exp[i(kx + ly - \omega t)]\,. \tag{3.1}$$

Substituting (3.1) into (2.7), it turns out that the free wave solutions satisfy the complex dispersion relation

$$\omega(k^2 + l^2 + a^{*2}) + \beta k + i\varkappa(k^2 + l^2) = 0\,, \tag{3.2}$$

where

$$a^{*2} = \frac{(f^2 + \varkappa^2)}{c^2}\,. \tag{3.3}$$

Equation (3.2) determines ω as a function of the wave numbers k and l in the form

$$\omega = \sigma(k, l) - i\gamma(k, l)\,, \tag{3.4}$$

where

$$\sigma \equiv -\frac{\beta k}{(k^2 + l^2 + a^{*2})} \quad \text{and} \quad \gamma = \frac{\varkappa(k^2 + l^2)}{(k^2 + l^2 + a^{*2})}\,. \tag{3.5a, b}$$

In contrast to the frequency of the free waves in an ocean with frictionless bottom, ω is complex with its real and imaginary parts giving the frequency and damping coefficients of the barotropic oscillations respecctively. The solution (3.1) represents the damping barotropic motions with the damping coefficient γ, and the westward and the southward phase velocities given by

$$-\frac{\sigma}{k} = \frac{\beta}{k^2 + l^2 + a^{*2}} = \frac{\beta}{m^2 + a^{*2}}\,, \tag{3.6}$$

$$-\frac{\sigma}{l} = \frac{\beta\left(\frac{k}{l}\right)}{k^2 + l^2 + a^{*2}} = \frac{\beta \cot\alpha}{m^2 + a^{*2}}\,, \tag{3.7}$$

where m and α represent the absolute wave number and the orientation of the waves defined by $(k, l) = (m\cos\alpha, m\sin\alpha)$. For a given latitude, if f and β are constants, the westward phase velocity depends only on the absolute wave number m, and the southward velocity depends on both m and α. Further, the components of the group velocity are

$$\left(\frac{\partial\sigma}{\partial k}, \frac{\partial\sigma}{\partial l}\right) = \frac{\beta}{(m^2 + a^{*2})^2}(m^2\cos 2\alpha - a^{*2},\ m^2\sin 2\alpha)\,. \tag{3.8a, b}$$

At a given latitude, the frequency $\sigma(k, l)$ is maximum when $(k, l) = (a^*, 0)$ and the maximum frequency $\sigma_{\max} = -\beta/2a^*$. This may be called the "cut-off" frequency of the wave motions and the magnitude of the cut-off frequency is smaller than that of the corresponding cut off frequency obtained by LONGUET-HIGGINS in his ocean model without bottom friction.

It may be recognized here that the characteristic features of the free wave solution (3.1) are modified by the frictional coefficient $\varkappa$.

4. The Forced Wave Solutions Due to Wind Stresses Periodic in Space

It is of interest to investigate the response of the ocean to the wind stress $\boldsymbol{\tau}$ which is harmonic in both space and time as

$$\boldsymbol{\tau} = \boldsymbol{\tau}_0 \exp[i(kx + ly - \omega' t)]\,, \tag{4.1}$$

where ω' is the prescribed frequency of the applied wind stress. In view of the fact that the component of $\boldsymbol{\tau}$ in the direction of the wave number has no contribution to the wind stress curl, we obtain

$$\mathrm{curl}_z\,\tau = \mathrm{im}\,\tau_0 \exp[i(kx + ly - \omega' t)]\,, \tag{4.2}$$

where τ_0 denotes the magnitude of the component of $\boldsymbol{\tau}_0$ parallel to the wave crests. This disturbance produces the flow field which is also in general harmonic in space and time as

$$v = V \exp\left[i(kx + ly - \omega' t)\right]. \tag{4.3}$$

The substitution of (4.3) into (2.7) gives the solution in the form

$$v = \frac{m\,\tau_0}{i\varrho h\beta} \cdot \frac{\sigma \exp\left[i(kx + ly - \omega' t)\right]}{\left[\omega' - (\sigma - i\gamma)\right]}. \tag{4.4}$$

The most distinctive feature of this solution is that it remains bounded for all ω' and indeed shows no resonant behaviour in the present ocean model. This conclusion is in contrast to that of LONGUET-HIGGINS in his frictionless model. Evidently, the small frictional force is capable of preventing the resonance.

We next consider an impulsive wind stress harmonic in space in the form

$$\operatorname{curl}_z \tau = \operatorname{im} \tau_0 \exp\{i(kx + ly)\}\, t_0 \delta(t), \tag{4.5}$$

where $\delta(t)$ is the DIRAC delta function and t_0 is the duration of the impulse which is small compared to the periodic time of the free wave motions (3.1). Using the improper contour integral representation of the DIRAC function as

$$\delta(t) = \frac{1}{2\pi} \int_C e^{-i\omega' t}\, d\omega', \tag{4.6}$$

where C is a contour from $-\infty$ to ∞ just above the real axis in the ω'-plane, the solution for v can be expressed as the contour integral

$$v = \frac{1}{2\pi i} \frac{m\tau_0 t_0}{\varrho h\beta} \int_C \frac{\sigma\, e^{i(kx+ly-\omega' t)}\, d\omega'}{(\omega' - \sigma + i\gamma)}. \tag{4.7}$$

The integrand of (4.7) has a simple pole at $\omega' = \sigma - i\gamma$ which lies on the third quadrant of the ω'-plane so that for $t < 0$, the contour C can be deformed into an infinite semi-circle in the upper-half plane and the integral along C is zero by CAUCHY's theorem. On the other hand, when $t > 0$, the integral can be deformed into a semi-circle in the lower-half plane enclosing the pole $\sigma - i\gamma$ so that the residue-contribution from this pole gives the solution as

$$v = -\frac{mQ\sigma}{\beta} \exp\left[i(kx + ly - \sigma t) - \gamma t\right], \quad t > 0 \tag{4.8}$$

which is, by (3.6),

$$v = -\frac{mkQ}{(m^2 + a^{*2})} \exp\left[i(kx + ly - \sigma t) - \gamma t\right], \tag{4.9}$$

where $Q = \tau_0 t_0/\varrho h$ represents the velocity that would be exerted on a fluid of layer of depth h by a uniformly distributed impulse $\tau_0 t_0$ acting on the fluid.

Under the assumptions made before, the divergence $u_x + v_y$ is small compared to the individual velocity gradients and hence a stream function ψ can be introduced, satisfying approximately,

$$\frac{\partial \psi}{\partial y} = -u, \quad \frac{\partial \psi}{\partial x} = v. \tag{4.10a, b}$$

Therefore ψ can readily be found from (4.9) and (4.10a) as

$$\psi = \frac{imQ}{(m^2 + a^{*2})} \exp\left[i(kx + ly - \sigma t) - \gamma t\right], \quad t > 0. \tag{4.11}$$

Thus the velocity vector (u, v) is directed to the wave crests and has magnitude

$$\frac{Qm^2}{(m^2 + a^{*2})} \exp\left[i(kx + ly - \sigma t) - \gamma t\right], \tag{4.12}$$

which at $t = 0$ is

$$\frac{Qm^2}{(m^2 + a^{*2})} e^{i(kx+ly)}. \tag{4.13}$$

If the disturbances were truely impulsive, the initial velocity at $t = 0$ should be $Q\, e^{i(kx+ly)}$ which differs from the calculated value (4.13) by an extra factor $m^2/(m^2 + a^{*2})$. This extra factor arises due to the adjustment of the motion to a quasi-geostrophic flow during the initial impulsive action.

5. The Transverse and the Longitudinal Wave Systems Due to an Arbitrary Wind Stress

We now consider an impulsive but arbitrary wind stress in the form

$$\operatorname{curl}_z \tau = \frac{\tau_0}{L} f(x, y)\, t_0 \delta(t)\,, \tag{5.1}$$

where L is a suitable length scale and $f(x, y)$ is an arbitrary function of x and y. On account of the FOURIER theorem, (5.1) can be written as

$$\operatorname{curl}_z \tau = \frac{\tau_0 t_0 \delta(t)}{L} \cdot \frac{L^2}{4\pi^2} \int_{-\infty}^{\infty} \int_{-\infty}^{\infty} \bar{f}(k, l)\, e^{i(kx+ly)}\, dk\, dl\,, \tag{5.2}$$

where

$$\bar{f}(k, l) = \frac{1}{L^2} \int_{-\infty}^{\infty} \int_{-\infty}^{\infty} f(x, y)\, e^{-i(kx+ly)}\, dx\, dy\,. \tag{5.3}$$

Then it follows from (4.5) and (4.11) that the stream function induced by the disturbance (5.2) is given by

$$\psi = \frac{QL}{4\pi^2} \int_{-\infty}^{\infty} \int_{-\infty}^{\infty} \frac{\bar{f}(k, l)}{(m^2 + a^{*2})} \exp\left[i(kx + ly - \omega t\right]\, dk\, dl\,. \tag{5.4}$$

In order to investigate the structure of the barotropic flow, it is necessary to evaluate (5.4). We next carry out the evaluation of (4.5) by the stationary phase method for large time t. The stationary points (k, l) are given by

$$\frac{\partial \sigma}{\partial k} = \frac{x}{t} \quad \text{and} \quad \frac{\partial \sigma}{\partial l} = \frac{y}{t}\,. \tag{5.5a, b}$$

Suppose that (k_n, l_n) represents one of the stationary points for a given x, y and t. Then the TAYLOR expansion of the phase function $(kx + ly - \sigma t)$ about the point (k_n, l_n) combined with the stationary phase condition (4.5a, b) leads us to the asymptotic evaluation (5.4) for large time t. Following the method of stationary phase as adopted by LONGUET-HIGGINS [11] or DEBNATH [12], it turns out that the stream function has the form, as $t \to \infty$,

$$\psi \sim \pm \frac{QL\bar{f}(k_n, l_n)}{\pi i t (m_n^2 + a^{*2}) \sqrt{|\Delta|}} \exp\left[i(k_n x + l_n y - \sigma_n t) - \gamma_n t\right] + 0\left(\frac{1}{t^2}\right), \quad \text{when } \Delta > 0 \tag{5.6}$$

according as $\sigma_{kk} > 0$ or < 0, where $\Delta = \sigma_{kk}\sigma_{ll} - \sigma_{kl}^2$, and σ_n, γ_n, m_n represent the values of σ, γ_1, m_n at (k_n, l_n) respectively,

$$\psi \sim \frac{QL\bar{f}(k_n, l_n)}{\pi t (m_n^2 + a^{*2}) \sqrt{|\Delta|}} \exp\left[i(k_n x + l_n y - \sigma_n t) - \gamma_n t\right] + 0\left(\frac{1}{t^2}\right), \quad \text{when } \Delta < 0\,. \tag{5.7}$$

It can be shown that the transverse system of waves is characterized by $\Delta > 0$ and the longitudinal wave system by $\Delta < 0$. Further, for the former system, σ_{kk} has the same sign as k. Hence, if the suffixes 1, 2 refer to the value of (k_n, l_n) for which $k_n > 0$ in the former and the latter wave systems, we obtain

$$\psi \sim \frac{QLa^*}{\pi t \beta} \left[A(k_1, l_1) \sin(k_1 x + l_1 y - \sigma_1 t) + A(k_2, l_2) \cos(k_2 x + l_2 y - \sigma_2 t)\right], \tag{5.8}$$

where

$$A(k, l) = \frac{\bar{f}(k, l)\, (m^2 + a^{*2})^{3/2}\, e^{-\gamma t}}{2a^* |(3k^2 - l^2)\, a^{*2} - m^4|^{1/2}}\,. \tag{5.9}$$

Evidently, the amplitudes of the above wave systems die out exponentially as $t \to \infty$. In view of the fact that $A(a^*, 0) = \exp(-\varkappa t/2)\, \bar{f}(a^*, 0)$, the stream function of the transverse wave system at the origin is

$$\psi_1 \sim \frac{QLa^*}{\pi t \beta}\, e^{-\frac{\varkappa t}{2}}\, \bar{f}(a^*, 0) \sin\left(a^* x + \frac{\beta t}{2a^*}\right). \tag{5.10}$$

This clearly indicates that the velocity of the transverse wave system is in the northward direction with exponentially damping amplitude.

The longitudinal wave systems, for which $m/a^* \to \infty$, are the non-divergent waves and are the limit of (5.8) as $a^*/m \to 0$. Thus the stream function for the wave systems is given by

$$\psi_2 \sim \frac{QLm}{2\pi \beta t}\, e^{-\varkappa t}\, \bar{f}(k_2, l_2) \cos(k_2 x + l_2 y - \sigma_2 t)\,. \tag{5.11}$$

In contrast to the solution for the longitudinal wave systems in an ocean without bottom friction, result (5.11) exhibits the pronounced effects of the frictional bottom and in turn the properties of the wave systems are modified.

Finally, the above analysis is in perfect agreement with that of Longuet-Higgins (11) as $\varkappa \to 0$.

6. Some Particular Wind Stress Curl

So far the pattern of the barotropic ocean currents has been investigated due to an impulsive atmospheric wind. Of special interest are some particular forms of $f(x, y)$ involved in the wind stress curl (5.1):

(a) $f(x, y) = \delta(x)\,\delta(y)$

(b) $f(x, y) = \exp(-r^2/4d)$, and

(c) $f(x, y) = H(d^2 - r^2),\ d > 0$,

where $H(x)$ is the Heaviside unit function and $r^2 = x^2 + y^2$.

Obviously, the wind stress curl related to case (a) represents the concentrated point disturbance whereas the wind stress curl associated with (b) and (c) corresponds to the disturbance distributed over a finite region in the x and y plane.

The function $\bar{f}(k, l)$ for the cases (a)—(c) is respectively given by 1, $\exp(-m^2 d)$ and $(d/m)\,J_1(md)$. These results enable us to find out the solution for the stream function and hence the velocity field in each of the above special cases.

It is noted that the principal features of the barotropic motions produced by the atmospheric winds (5.1) with the special cases (a)—(c) would be similar to those described in the preceeding section. Hence any further reproduction of the results and their discussion would be redundant.

7. Barotropic Response to Travelling Storms

It is of special interest to investigate the long period barotropic response of the ocean model to an impulsive wind stress. We consider a circular storm travelling rapidly with a uniform speed V on a straight course along the northern direction so that at $t = 0$ the wind stress curl has the form

$$\operatorname{curl}_z \tau = \frac{\tau_0}{L} f(x, y)\,. \tag{7.1}$$

Replacing y by Vt in (7.1) and integrating with respect to time, the total impulsive wind stress exerted on the fluid is given by

$$\int_{-\infty}^{\infty} \operatorname{curl}_z \tau \, dt = \frac{\tau_0}{L} \int_{-\infty}^{\infty} f(x, Vt)\, dt = \frac{\tau_0 t_0}{L^2} \int_{-\infty}^{\infty} f(x, y)\, dy, \qquad L = V t_0\,, \tag{7.2}$$

which is evidently equivalent to an impulse given by

$$\operatorname{curl}_z \tau = \tau_0 f(x)\, t_0 \delta(t)\,, \tag{7.3}$$

provided that the time integral of (7.3) is equal to (7.2), that is, if

$$f(x) = \frac{1}{L^2} \int_{-\infty}^{\infty} f(x, y)\, dy\,, \tag{7.4}$$

which is, by the Fourier integral representation of $f(x, y)$,

$$f(x) = \frac{1}{L^2}\frac{L^2}{4\pi^2} \int_{-\infty}^{\infty} \left[\int_{-\infty}^{\infty}\int_{-\infty}^{\infty} \bar{f}(k, l)\, e^{i(kx+ly)}\, dk\, dl \right] dy\,. \tag{7.5}$$

In view of the Fourier theorem, we obtain

$$f(x) = \frac{1}{2\pi} \int_{-\infty}^{\infty} \bar{f}(k, 0)\, e^{ikx}\, dk\,, \tag{7.6}$$

so that the impuls (7.3) becomes

$$\operatorname{curl}_z \tau = \frac{\tau_0 t_0 \delta(t)}{2\pi} \int_{-\infty}^{\infty} \bar{f}(k, 0)\, e^{ikx}\, dk\,. \tag{7.7}$$

Hence the stream function for the barotropic response to the above impulse has the form

$$\Psi = \frac{Q}{2\pi} \int_{-\infty}^{\infty} \frac{\bar{f}(k, 0)}{(k^2 + a^{*2})} \exp\left[i(kx - \omega t)\right] \mathrm{d}k\,, \tag{7.8}$$

where $\omega(k) \equiv \sigma(k) - i\gamma(k)$, $\sigma(k)$ and $\gamma(k)$ are given by (3.5a, b) with $l = 0$.

We apply the stationary phase method for the asymptotic evaluation of (7.8). It is evident that all possible waves at a given point (x, y) are determined by the real stationary points of (7.8), that is, the real solutions of the equation

$$\frac{\mathrm{d}\sigma}{\mathrm{d}k} = \frac{x}{t}. \tag{7.9}$$

Introducing the notations $\zeta = xa^{*2}/\beta t$ and $\eta = k/a^*$, equation (7.9) can be written as

$$\zeta = \frac{a^{*2}}{\beta} \frac{\mathrm{d}\sigma}{\mathrm{d}k} = \frac{\eta^2 - 1}{(\eta^2 + 1)^2}, \tag{7.10}$$

so that

$$\eta^2 = \frac{1}{2\zeta}\left[(1 - 2\zeta) \pm (1 - 8\zeta)^{1/2}\right]. \tag{7.11}$$

It is noted that $(a^{*2}/\beta)\,(\mathrm{d}\sigma/\mathrm{d}k)$ has a minimum value -1 at $\eta = 0$ and has maximum at $\eta = \pm\sqrt{3}$. The graph of $(a^{*2}/\beta)\,(\mathrm{d}\sigma/\mathrm{d}k)$ against η is shown in Figure 1, which indicates that (7.10) has solutions provided ζ lies in the domain $-1 \leqq \zeta \leqq 1/8$. More precisely, there exists one system of waves in $-1 \leqq \zeta < 0$ and two systems of waves in $0 < \zeta \leqq 1/8$.

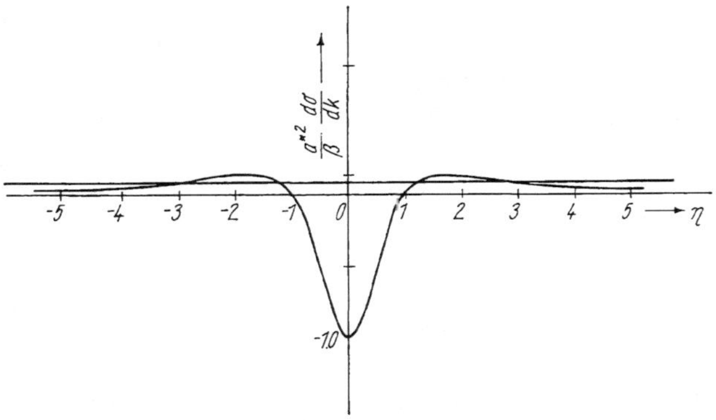

Fig. 1. The group velocity as a function of wave number

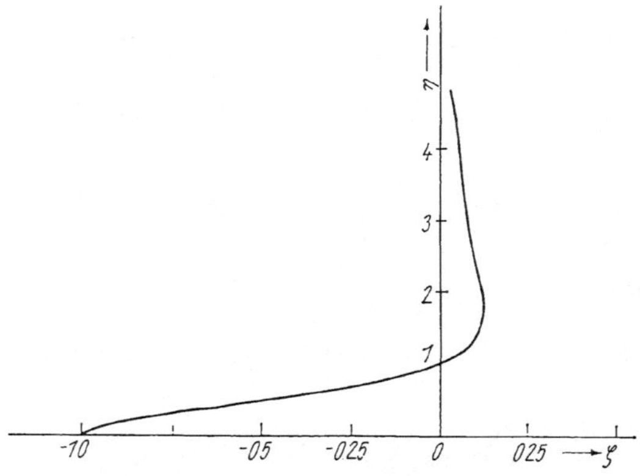

Fig. 2. The wave number η as a function of ζ

The wave number η as a function of ζ given in (7.11) is also represented by Figure 2. It is also of interest to note that η increases steadily from 0 at $\zeta = -1$ to $\sqrt{3}$ at $\zeta = 1/8$ with $\eta = 1$ at $\zeta = 0$ which corresponds to the first system of waves. Then $\eta(\zeta)$ continues to increase monotonically from $\zeta \geqq 1/8$ with the fact that $|\eta| \to \infty$ as $\zeta \to 0+$. Thus the second system of waves is a continuation of the first in $0 < \zeta \leqq 1/8$.

Thus the stationary phase approximation for large time gives the asymptotic solution for the stream function (7.8) in the form

$$\Psi \sim \frac{Q}{2\pi} \sum_{k=k_n} \frac{\bar{f}(k, 0)\, \mathrm{e}^{-t\gamma(k)}}{(k^2 + a^{*2})} \left\{\frac{2\pi}{it|\sigma''(k)|}\right\}^{1/2} \exp\left[i(kx - \sigma t)\right], \tag{7.12}$$

where

$$\sigma''(k) \equiv \frac{\mathrm{d}^2\sigma}{\mathrm{d}k^2} = 2\beta k \frac{(3a^{*2} - k^2)}{(k^2 + a^2)^3}, \tag{7.13}$$

and the summation is extended over all the points of intersection depicted in Figure 1.

If k_1 and k_2 represent the first two positive roots of (7.11), then the stream function reduces to the asymptotic form

$$\Psi \sim \frac{Q}{(\pi a^* \beta t)^{1/2}} \left[\bar{f}(k_1, 0)\, |A(k_1)|\, \mathrm{e}^{-t\gamma(k_1)} \cos\left(k_1 x - \sigma_1 t - \frac{\pi}{4}\right) + \right.$$

$$\left. + \bar{f}(k_2, 0)\, |A(k_2)|\, \mathrm{e}^{-t\gamma(k_2)} \cos\left(k_2 x - \sigma_2 t + \frac{\pi}{4}\right)\right], \tag{7.14}$$

where

$$A(k) = \left[\frac{a^*(k^2 + a^{*2})}{k(k^2 - 3a^{*2})}\right]^{1/2}. \tag{7.15}$$

Because of the fact that $A(k) \to \infty$ when $k = 0$ and $k = a^*\sqrt{3}$, the above solution remains valid except at $k = 0$ and $k = a^* \sqrt{3}$. It is clear that the points $k = 0(\zeta = -1)$ and $k = a^* \sqrt{3}$ $(\zeta = 1/8)$ render the stationary phase approximation invalid. It is then necessary to work out the solution in the neighborhood of these points using the higher order approximation of the stationary phase method. We shall return to this point in the next section.

It is worth noting that the bottom friction introduces an additional exponential damping factor $\exp[-t\gamma(k)]$ in the amplitude of the two waves given obtained in (7.14). The velocity components $(\partial\Psi/\partial y, -\partial\Psi/\partial x)$ of these waves clearly ensure that the velocity of both the waves is in a northern direction and has magnitude

$$V = \frac{Q}{\sqrt{\pi}}\left(\frac{a^*}{\beta t}\right)^{1/2} \bar{f}(k_r, 0)\, |A(k_r)| \left(\frac{k_r}{a^*}\right) \exp[-t\gamma(k_r)], \tag{7.16}$$

where $r = 1, 2$. In particular, (7.16) represents the velocity for the concentrated point disturbance or the distributed disturbance according as $\bar{f}(k, 0)$ is 1 or $\exp(-m^2 d)$ and $d/m\, J_1(md)$.

It is of interest to consider the case of a circular storm in which the radial component of the wind speed is zero and the transverse component is non-zero so that the curl of the wind stress τ_θ is given by

$$\mathrm{curl}_z \tau = \frac{1}{r}\frac{\partial}{\partial r}(r\tau_\theta), \tag{7.17}$$

where

$$\tau_\theta = \tau_0 \left(\frac{r}{r_0}\right)^2 \exp\left(-\frac{r}{r_0}\right) t_0 \delta(t). \tag{7.18}$$

Thus the Fourier transform of the function $f(x, y)$ related to (7.17) is given by

$$\bar{f}(k, l) = \frac{1}{r_0^2}\int_{-\infty}^{\infty}\int_{-\infty}^{\infty} f(x, y)\, e^{-i(kx+ly)}\, dx\, dy, \tag{7.19}$$

where

$$f(x, y) = \left(\frac{L}{r_0}\right)\left[3\left(\frac{r}{r_0}\right) - \left(\frac{r}{r_0}\right)^2\right] \exp\left(-\frac{r}{r_0}\right), \tag{7.20}$$

with $L = r_0$.

Invoking the polar coordinates in both (x, y) and (k, l) planes and integrating with respect to θ along with the result 8.2 (20) of ERDELYI [12], it turns out that

$$\bar{f}(k, l) = \frac{6\pi r_0^2 m^2(4 - m^2 r_0^2)}{(1 + m^2 r_0^2)^{7/2}}, \tag{7.21}$$

so that

$$\bar{f}(k, 0) = \frac{6\pi r_0^2 k^2(4 - k^2 r_0^2)}{(1 + k^2 r_0^2)^{7/2}}. \tag{7.22}$$

Substituting (7.22) into the expression (7.16), we can determine the velocity associated with the circular storm

$$V = \frac{Q}{\sqrt{\pi}}\left(\frac{a^*}{\beta t}\right)^{1/2} e^{-t\gamma(k)} \left(\frac{k}{a^*}\right) |A(k)| \left[\frac{6\pi r_0^2 k^2(4 - k^2 r_0^2)}{(1 + k^2 r_0^2)^{7/2}}\right]. \tag{7.23}$$

This shows that the velocity depends on the dimension r_0 of the storm. At a fixed time, when $kr_0 \to \infty$, the factor $\bar{f}(k, 0)$ in (7.23) enables to make the velocity zero.

8. Higher Order Asymptotic Approximation when $\sigma''(k)$ Vanishes

Making reference to the standard result (COPSON, [13]) for higher order approximation in terms of the AIRY integral involved in the stationary phase method, the asymptotic solution for the stream function given by (7.8) can be obtained as

$$\Psi \sim \frac{Q\bar{f}(k, 0)}{(k^2 + a^{*2})}\left(\frac{-2}{t\sigma'''(k)}\right)^{1/3} Ai\left[\left(\frac{-2}{t\sigma'''}\right)^{1/3}(x - \sigma' t)\right] e^{i(kx - \omega t)}, \tag{8.1}$$

where the value of k corresponds to $\sigma''(k) = 0$, $Ai(z)$ represents the AIRY function defined by

$$Ai(z) = \frac{1}{2\pi} \int_{-\infty}^{\infty} e^{i\left(\frac{1}{3}t^3 + zt\right)} dt\,, \tag{8.2}$$

and

$$\sigma'''(k) = \frac{6\beta(k^4 + a^{*4} - 6a^{*2}k^2)}{(k^2 + a^{*2})^4}. \tag{8.3}$$

In the neighborhood of $k = 0$, $\sigma'''(k) = 6\beta/a^{*4}$, solution (8.1) reduces to the form

$$\Psi \sim -\left(\frac{8}{3}\right)^{1/3} Q\,\bar{f}(0, 0) \left(\frac{1}{a^*\beta t}\right)^{1/3} e^{-t\gamma(0)} Ai\left[\left(-\frac{a^{*4}}{3\beta t}\right)^{1/3} \left(x + \frac{\beta t}{8a^{*2}}\right)\right]. \tag{8.4}$$

Similarly, the solution for the stream function in the neighborhood of $k = a^* \sqrt{3}$ is given by

$$\Psi \sim \left(\frac{4}{3}\right)^{1/3} Q\bar{f}\,(\sqrt{3}\,a^*, 0) \left(\frac{1}{a^*\beta t}\right)^{1/3} e^{-t\gamma(a^*\sqrt{3})}\, Ai\left[\left(-\frac{32a^{*4}}{3\beta t}\right)^{1/3} \left(x - \frac{\beta t}{8a^{*2}}\right)\right]. \tag{8.5}$$

The effect of the bottom friction is reflected in the above results and the motion decays eventually as $t \to \infty$. This conclusion shows a striking contrast to that of LONGUET-HIGGINS.

9. Concluding Remarks

The present analysis has been carried out for barotropic oceanic motion and can be generalized to baroclinic motions in a stably stratified ocean. The study of the baroclinic motions is especially important for an understanding of the catastropic effects of an intense hurricane, violent storm, underwater earthquake or explosion, etc. because a major part of the input energy involved in any one of the above disturbances is expended to produce internal .or baroclinic modes of oscillation. Such study is underway and will be presented in a subsequent paper.

In this connection, mention may be made of the initial value investigation of DEBNATH [14] into the generation and propagation of internal wave motions in two-layered stably stratified non-rotating oceans due to oscillating pressure distributions. Among many other conclusions, this analysis had indicated the existence of violent internal wave motions for small values of the density parameter. In recent papers, DEBNATH in collaboration with KULCHAR [15] and BAGCHI [16] has presented a linearized theory of dispersive unsteady long waves on a rotating ocean created by stationary or traveling but arbitrary wind stress distributions which may be steady, oscillatory or transient in character acting on the free surface of the rotating ocean. The solution for the steady-state as well as the transient wave motion has been explicitly obtained by means of the generalized FOURIER analysis. It has been shown that the transient solution eventually decays to zero as $t \to \infty$ and the steady-state wave motion is reached. The nature of the steady-state motion has been investigated for different values of frequency parameter ω/f and of the velocity parameter U/c where ω, U are the frequency and the velocity of the disturbance, f is the CORIOLIS parameter and $c = \sqrt{gh}$ is the phase velocity of the non-dispersive shallow water waves in a non-rotating ocean.

It has already been mentioned that the present study can be used as a guide for an investigation of the non-linear frictional ocean model with or without viscous effects. This work seems to be useful for practical applications in the dynamics of ocean and hence will be undertaken in a future project.

References

1 ROSSBY, C. G., On the Mutual Adjustment of Pressure and Velocity Distribution in Certain Simple Current Systems, II, J. Mar. Res. **1** (1938), 239–263.
2 ROSSBY, C. G., On the Dispersion of Planetary Waves in a Barotropic Atmosphere, Tellus **1** (1949), 54–58.
3 SVERDRUP, H. U., Wind-driven Currents in a Baroclinic Ocean with Application of the Equatorial Currents of the Eastern Pacific, Proc. Nat. Acad. Sc. **33** (1947), 318–326.
4 STOMMEL, H., The Westward Intensification of Wind-driven Ocean Currents, Trans. Amer. Geophys. Un. **29** (1948), 202–206.
5 CHARNEY, J. G., Generation of Ocean Currents by Winds, J. Mar. Res. **14** (1955), 477–498.
6 BOLIN, B., The Adjustment of a non-balanced Velocity Field Towards Geostrophic Equilibrium in a Stratified Fluid, Tellus **5** (1953), 273–285.
7 CAHN, A., An Investigation of the Free Oscillations In a Simple Current Systems, J. Meteorol. **2** (1945), 113–119.
8 VERONIS, G. and STOMMEL, H., The Action of Variable Wind Stresses on a Stratified Ocean, J. Mar. Res. **14** (1955), 43–75.
9 VERONIS, G., Partition of Energy between Geostrophic and Non-Geostrophic Oceanic Motions, Deep Sea Res. **3** (1956) 157–177.
10 VERONIS, G., On the Transient Response of a β-plane ocean, J. Oceanogr. Soc. Japan **14** (1958), 3–5.
11 LONGUET-HIGGINS, M. S., The Response of a Stratified Ocean to Stationary or Moving Wind-Systems, Deep Sea Res. **12** (1965), 923–973.
12 ERDELYI, A., Tables of Integral Transforms, McGraw-Hill, New York (1954).
13 COPSON, E. T., Asymptotic Expansions, Cambridge University Press (1965).

14 DEBNATH, L., On Internal Wave Motions in Stratified Oceans, Indian J. Pure and Applied Math. 1 (1970), 372–389.
15 DEBNATH, L. and KULCHAR, A. G., On the Generation of Dispersive Long Waves on a Rotating Ocean by Wind Stress Distributions, J. Phys. Soc. Japan 33 (1972), 1464–1470.
16 BAGCHI, K. K. and DEBNATH, L., Long Waves on a Rotating Sea Due to Traveling Disturbances, J. Phys. Soc. Japan 35 (1973), 281–285.

Eingereicht am 2. 7. 1975

Anschrift: Dr. LOKENATH DEBNATH, Department of Mathematics, East Carolina University, Greenville, North Carolina 27834, USA

ZAMM 56, 479 – 489 (1976)

N. Liron / H. E. Wilhelm

Similarity Solutions for Magnetogasdynamic Outflows and Inflows with Viscous, Ohmic, and Thermal Dissipation between Inclined Walls*

Es werden numerische Ergebnisse für Ein- und Ausströmungen eines kompressiblen, ionisierten Gases zwischen ebenen geneigten Wänden unter Berücksichtigung der Lorentzkräfte, hervorgerufen durch ein azimutales Magnetfeld, des viskosen Impulstransportes und der thermischen, ohmschen sowie viskosen Energiedissipation angegeben. Die Analyse basiert auf einer Ähnlichkeitstransformation des Strömungsfeldes, wodurch sich die (nichtlinearen) partiellen Differentialgleichungen der Magnetogasdynamik auf gewöhnliche reduzieren. Außer den Lösungen einer normalen, monotonen Aus- und Einströmung existieren Lösungen für oszillierende Ausströmung und „gemischte" Aus- und Einströmung im stationären Zustand für gewisse Werte der Reynolds-, Prandtl-, Mach- und Hartmann-Zahlen. Mathematische Theoreme, die die Bedingungen für das Auftreten derartiger Strömungszustände definieren, werden aufgestellt.

Numerical solutions are presented for outflows and inflows of a compressible, ionized gas between plane inclined walls, considering Lorentz forces due to an azimuthal magnetic field, viscous momentum transfer, thermal, ohmic, and viscous energy dissipation. The analysis is based on a similarity transformation of the flow fields which reduces the (nonlinear) partial differential equations of magnetogasdynamics into ordinary ones. In addition to regular, monotone outflow and inflow solutions, "oscillatory" outflow and "mixed" outflow and inflow solutions are shown to exist in steady state for certain values of the Reynolds, Prandtl, Mach, and Hartmann numbers. Mathematical theorems are established which define the conditions for the occurrence of these flow types.

Даются численные результаты в случае втекания и вытекания некоторого сжимаемого ионизированного газа между плоскими наклонёнными стенками с учётом сил Лоренца. Эти течения вызываются посредством действия азимутального магнитного поля, вязкого переноса импульса и термической, вязкой диссипации энергии, а также диссипации энергии Ома. Анализ базируется на некотором преобразовании подобия поля скоростей, с помощью которого дифференциальные уравнения в частных производных (нелинейные) магнетогазодинамики превращаются в обыкновенные дифференциальные уравнения. Исключая решения для нормального монотонного вытекания и втекания, а также решения для осциллирующего вытекания и „смешанного" вытекания и втекания показывается, что стационарное состояние существует для определённых значений чисел Рейнольдса, Прандтля, Маха и Хартманна. Строятся математическкие теории, которые определяют условия, при которых наступают названные состояния течения.

Introduction

Nonlinear partial differential equations can frequently be reduced to ordinary nonlinear differential equations by means of similarity transformations [1]. By this method, the first exact solutions of the Navier-Stokes equations have been found by Hamel [2] and Landau [3] for isothermal plane diffuser and conical jet flows, respectively. In another classical work, Millsaps and Pohlhausen gave a closed-form similarity solution for incompressible flow with heat transport between inclined walls [4]. More recently, similarity transformations have been found for compressible magnetohydrodynamic flows between diverging walls by Wilhelm [5, 6]. In the polytropic case, the resulting nonlinear eigenvalue problem for ordinary differential equations [5] has been solved analytically in an approximation based on large Reynolds numbers by Wilhelm [5] and numerically for arbitrary (small or large) Reynolds numbers by Liron and Wilhelm [7]. Based on a similarity transformation [6], exact numerical solutions are presented in the following for the flow of a compressible, viscous, electrically and thermally conducting gas (dissipation by viscous momentum transfer, heat conduction, ohmic and viscous heating) in plane diffusers and confusers with an external magnetic field in the azimuthal direction. As an analytical contribution, the conditions for the existence of pure inflows and outflows and mixed flows are discussed.

As to further applications of similarity theory, such as explosion and implosion problems in gasdynamics, it is referred to Sedov [8] and Stanyukovich [9]. It is remarkable that other methods of solution for nonlinear partial differential equations may also result in similarity solutions, e.g., in electrohydrodynamics (space charge flow) the von Mises transformation leads to a functional similarity solution [10].

Theoretical Foundations

Let cylindrical coordinates $(\bar{r}, \theta, \bar{z})$ be introduced for the description of the magnetogasdynamic diffuser or confuser (Fig. 1). The ionized gas flow is bounded in the planes $(\theta = \pm\, \theta_0, \bar{r}_1 \leqq \bar{r} \leqq \bar{r}_2)$ by insulating walls, and quasi unbounded in the directions parallel to the $\bar{z}$ axis. The latter assumption is applicable to a finite duct with electrode plates at $\bar{z} = \pm\, \bar{z}_\infty$, where $\bar{z}_\infty \gg 1/2\, (\bar{r}_1 + \bar{r}_2)\, \theta_0$. The magnetic field has its sources in an electric current I flowing through a conducting rod $(0 \leqq \bar{r} \leqq \bar{r}_a,\ -\infty < \bar{z} < +\infty,\ \bar{r}_a < \bar{r}_1)$, and is given by ($\mu_0$ = permeability of vacuum)

$$\overline{B}_\theta = \frac{\mu_0}{2\pi} \frac{I}{\bar{r}}, \qquad \bar{r}_a \leqq \bar{r} < \infty .$$

*) Supported in part by the US Office of Naval Research.

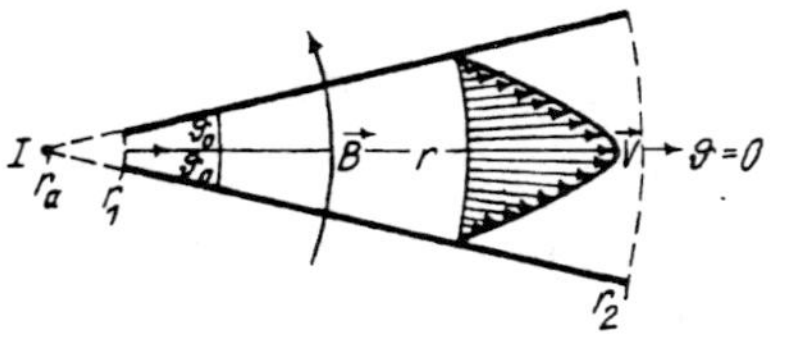

Fig. 1. Geometry of duct with azimuthal magnetic field $\vec{B}$ and velocity distribution $\vec{V}$ (outflow)

In the absence of flow sources or sinks at the inclined walls, the velocity field is radial, $\bar{v}_r = \bar{u}$, if the force fields are radial. The flow of the plasma (conductivity σ) across the magnetic field induces a current density field which is axial, $\bar{j}_z = \sigma(\bar{E}_z + \bar{u}\bar{B}_\theta)$, if (i) the induced magnetic fields and (ii) the Hall effect are negligible i.e.,

$$\text{(i)} \quad R_B = \mu_0 \sigma \bar{u}(\bar{r}, \theta)\, \bar{r} \ll 1\,, \qquad \bar{r}_1 \leqq \bar{r} \leqq \bar{r}_2\,, \qquad |\theta| \leqq \theta_0\,,$$

$$\text{(ii)} \quad \omega\tau = (\sigma/n\,|e|)\, \bar{B}_\theta \ll 1\,.$$

The resulting Lorentz force density then is a purely radial field which opposes the inducting flow, $(\bar{j} \times \bar{B})_r = -\sigma(\bar{E}_z + \bar{u}\bar{B}_\theta)\,\bar{B}_\theta$. Since $\nabla \times \bar{E} = 0$ and $\nabla \cdot \bar{j} = 0$, one has

$$\bar{E}_z = 0 \quad \text{for} \quad \bar{E}_{\bar{z} = \pm \bar{z}_\infty} = 0\,.$$

The axial current density $\bar{j}_z$ flows through the planes $\bar{z} = \pm\, \bar{z}_\infty$ (electrodes) and generates an electrical current J in the external circuit,

$$J = I\, \frac{\mu_0}{2\pi}\, \sigma \int\limits_{\bar{r}_1}^{\bar{r}_2} \int\limits_{-\theta_0}^{+\theta_0} \bar{u}(\bar{r}, \theta)\, d\bar{r}\, d\theta\,.$$

The magnetogasdynamic flow model under consideration is described by the nonlinear boundary-value problem for the dimensionless velocity $[u = u(r, \theta)]$, density $[\varrho = \varrho(r, \theta)]$, and pressure $[p = p(r, \theta)]$ fields:

$$\varrho u \frac{\partial u}{\partial r} = -\frac{1}{\gamma M^2}\frac{\partial p}{\partial r} - \frac{H^2}{R}\frac{u}{r^2} + \frac{4}{3R}\frac{\partial}{\partial r}\left(\frac{1}{r}\frac{\partial}{\partial r}(ur)\right) + \frac{1}{R}\frac{\partial^2 u}{r^2 \partial\theta^2}\,, \tag{1}$$

$$0 = -\frac{1}{\gamma M^2}\frac{\partial p}{r\,\partial\theta} + \frac{1}{3R}\frac{\partial}{r\,\partial\theta}\left(\frac{1}{r}\frac{\partial}{\partial r}(ru)\right) + \frac{2}{R}\frac{1}{r^2}\frac{\partial u}{\partial\theta}\,, \qquad \frac{\partial}{\partial r}(\varrho u) + \frac{\varrho u}{r} = 0\,, \tag{2) (3}$$

$$\frac{1}{r}\frac{\partial}{\partial r}(\varrho r T u) = \frac{\gamma}{PR}\left(\frac{\partial^2 T}{\partial^2 r^2} + \frac{1}{r}\frac{\partial T}{\partial r} + \frac{1}{r^2}\frac{\partial^2 T}{\partial\theta^2}\right) - (\gamma - 1)\, p\left(\frac{\partial u}{\partial r} + \frac{u}{r}\right) + \frac{\gamma(\gamma-1)\, M^2 H^2}{R}\left(\frac{u}{r}\right)^2 -$$
$$- \frac{2}{3}\frac{\gamma(\gamma-1)\, M^2}{R}\left(\frac{1}{r}\frac{\partial}{\partial r}(ru)\right)^2 + \frac{\gamma(\gamma-1)\, M^2}{R}\left[2\left(\frac{\partial u}{\partial r}\right)^2 + \left(\frac{\partial u}{r\,\partial\theta}\right)^2 + 2\left(\frac{u}{r}\right)^2\right], \tag{4}$$

$$p = \varrho T\,, \tag{5}$$

where

$$u(r, \theta = \pm\,\theta_0) = 0\,, \quad r_1 \leqq r \leqq r_2\,, \qquad T(r, \theta = \pm\,\theta_0) = T_\pm(r)\,, \quad r_1 \leqq r \leqq r_2\,, \tag{6) (7}$$

and

$$Q = \int_{-\theta_0}^{+\theta_0} \varrho u r\, d\theta\,. \tag{8}$$

The conservation equations for momentum [Eqs. (1)—(2)], mass [Eq. (3)], and energy [Eq. (4)] are subject to the boundary conditions in Eqs. (6)—(8) which specify that at the walls i) the velocity field vanishes, ii) the temperature field assumes a prescribed distributions $T_\pm(r)$, and iii) the mass flow rate per unit height $\Delta z = 1$ is an invariant, Q. Equation (8) is mathematically equivalent to specifying "selfsimilar" boundary conditions, $u(r = r_{1,2}, \theta) = r_{1,2}{}^{-m} g(\theta)$ at the entrance and exit boundaries ($r = r_{1,2}$, $|\theta| \leqq \theta_0$). Equation (5) is the equation of state.

The Reynolds (R), Mach (M), Prandtl (P), Hartmann (H) numbers and the ratio of specific heats (γ) are defined by

$$R = \bar{\varrho}_0 \bar{u}_0 \bar{r}_0/\mu\,, \quad M = \bar{u}_0/(\gamma \bar{p}_0/\bar{\varrho}_0)^{1/2}\,, \quad P = \mu c_p/\lambda\,, \quad H = (\sigma/\mu)^{1/2}\, \bar{B}_\theta(\bar{r}_0)\, \bar{r}_0\,,$$
$$\gamma = c_p/c_v = (1 + \mathcal{R}/c_v)\,, \tag{9}$$

where

$$\bar{\varrho}_0 = \bar{\varrho}(\bar{r}_0, \hat{\theta}) > 0\,, \quad \bar{u}_0 = \bar{u}(\bar{r}_0, \hat{\theta}) \neq 0\,, \quad \bar{p}_0 = \bar{p}(\bar{r}_0, \hat{\theta}) > 0\,, \quad T_0 = T(\bar{r}_0, \hat{\theta}) > 0\,, \quad |\hat{\theta}| < \theta_0\,, \tag{10}$$

are known reference values of the (corresponding dimensional) flow fields, e.g., for $\bar{r}_0 = \bar{r}_1$ and $\hat{\theta} = 0$:

$$\bar{\varrho}_0 = \bar{\varrho}(\bar{r}_1, 0)\,, \qquad \bar{u}_0 = \bar{u}(\bar{r}_1, 0)\,, \qquad \bar{p}_0 = \bar{p}(\bar{r}_1, 0)\,, \qquad \bar{T}_0 = \bar{T}(\bar{r}_1, 0)\,. \tag{11}$$

It should be noted that the viscosity (μ), electrical (σ) and thermal (λ) conductivities are assumed to be constant. Diffuser ($>$) and confuser ($<$) flows are characterized by different signs of the net flow rate,

$$Q \gtrless 0 .$$

Nonlinear Eigenvalue Problem

The boundary-value problem for nonlinear, partial differential equations [Eqs. (1)—(8)] is reduced by means of the similarity transformation [6],

$$u(r, \theta) = r^{-m} g(\theta) , \qquad p(r, \theta) = r^{-(m+1)} h(\theta) , \tag{12) (13}$$

$$\varrho(r, \theta) = r^{m-1} f(\theta)/g(\theta) , \qquad T(r, \theta) = r^{-2m} \varphi(\theta) , \tag{14) (15}$$

where

$$\varphi(\theta) \equiv \frac{g(\theta)\, h(\theta)}{f(\theta)}, \qquad h(\theta) = 1 + \frac{7-m}{3} \gamma \frac{M^2}{R} [g(\theta) - 1] , \tag{16) (17}$$

to an ordinary nonlinear eigenvalue problem for the similarity parameter m:

$$g'' + [(m+1)^2 - H^2]\, g + mRg^2 (1 - s + sg)\, \varphi^{-1} + \frac{1}{3} (m+1)(1-s) \frac{7-m}{s} = 0 , \tag{18}$$

$$\varphi'' + (2m)^2 \varphi = - PEF\big(g(\theta), g'(\theta)\big) , \tag{19}$$

with

$$F(g, g') \equiv (g')^2 + \left[\frac{4}{3}(m^2 + m + 1) + H^2\right] g^2 + \frac{R}{\gamma E} [2m - (1-m)(\gamma - 1)] (1 - s + sg)\, g , \tag{20}$$

where

$$g(\theta = \pm \theta_0) = 0 , \qquad \varphi(\theta = \pm \theta_0) = \varphi_\pm . \tag{21) (22}$$

As a result of the normalization [Eq. (10)], the θ-dependent functions have to satisfy, in addition, the conditions,

$$g(\hat{\theta}) = 1, \dots , \varphi(\hat{\theta}) = 1 . \tag{23}$$

In Eqs. (18)—(20), the abbreviations E and s have been introduced, which do not represent independent flow parameters but rather combinations of (γ, M) and (m, γ, M, R), respectively,

$$E = (\gamma - 1)\, M^2 , \quad \text{with} \quad \Gamma \equiv \gamma M^2 = \frac{c_p E}{\mathcal{R}} = \frac{\gamma}{\gamma - 1} E , \qquad s = \frac{7-m}{3} \frac{\gamma M^2}{R} . \tag{24) (25}$$

Once $g(\theta)$ and $\varphi(\theta)$ are determined from the boundary-value problem in Eqs. (18)—(22), $h(\theta)$ and $f(\theta)$ follow directly from Eqs. (16) and (17), respectively. Since the (absolute) temperature of a gas cannot be zero, one has $\varphi(\theta) > 0$ for any $|\theta| \leqq \theta_0$ in Eq. (18). Since $p(r, \theta) \geqq 0$ and $T(r, \theta) > 0$, the restrictions

$$h(\theta) \geqq 0 , \qquad \varphi(\theta) > 0 , \tag{26) (27}$$

are imposed on the solutions for physical reasons. Combining Eqs. (17) and (25) gives

$$h(\theta) = sg(\theta) + 1 - s , \tag{28}$$

whence

$$s \leqq 1 , \tag{29}$$

by Eq. (26) since $g(\theta_0) = 0$. More generally, Eq. (26) takes the form

$$g(\theta) \geqq - \frac{1-s}{s} , \qquad s > 0 , \qquad g(\theta) \leqq \frac{1+s'}{s'} , \qquad s' = -s > 0 . \tag{30}$$

Looking at the transformation (12)—(17), it is seen that the parameter m has four different regions:

(1) $m > 1$, for which u, p and T decrease with r but ϱ increases;

(2) $0 < m < 1$, for which u, p, T and ϱ decrease with r;

(3) $-1 < m < 0$, for which u and T increase with r and p, ϱ decrease with r;

(4) $m < -1$, for which u, T and p increase with r but ϱ decreases. One would expect differences in behavior of the solution in the four regions.

For an incompressible ($\gamma = \infty$), isothermal ($\varphi = 1$), nonconducting ($H = 0$) fluid, (18) reduces to the (non-dimensional) differential equation of the classical HAMEL flow [2]—[4] since $s \to 2u_0^2(p_0/\varrho_0)\, R \neq 0$ for $\gamma \to \infty$ by Equations (25) and (9).

Numerical Solutions for Outflows, Inflows, and Mixed Flows

The eigen-value m is determined by the condition for the existence of a solution of the boundary-value problem [Eqs. (18)—(22)]. This is accomplished by numerical iteration. In the lowest approximation, the eigen-value is assumed to be given by its value in the polytropic approximation [6], $m = (k-1)/(k+1)$, where k is the polytropic coefficient, $k \cong \gamma$. Since $\varphi(\theta = 0) = 1$ and $g(\theta = 0) = 1$ (nondimensionalization) and $\varphi'(\theta = 0) = 0$ and $g'(\theta = 0) = 0$ are known from the symmetry properties and boundary conditions of the flow, the differential equations [Eqs. (18)—(19)] can be integrated in analogy to an initial-value problem (instead of a 2-point boundary-value problem). The first value $\theta = \theta_0$, for which the velocity field becomes zero, $g(\theta = \theta_0) = 0$, gives the location $\theta = \theta_0$ of the wall, associated with the particular eigen-value m. In the following approximations, the eigen-value m is increased or decreased in small steps Δm until the numerical $g(\theta)$ — curve becomes zero at the desired value $\theta = \theta_0$. In a similar way, one integrates the differential equations in the case of asymmetrical boundary conditions for the temperature field $\varphi(\theta)$, i.e., in the case of asymmetrical flow fields $\varphi(\theta)$ and $g(\theta)$. In this case, the "initial-value" problem has to be integrated for various values assigned to $\varphi'(\theta = 0)$ until the desired boundary-values are found for $\varphi(\theta)$ at the walls.

In partially and in particular in weakly ionized gases, viscous momentum transfer in the flow is determined by atom-atom (00) collisions, whereas heat conduction is mainly due to electron-electron (ee) collisions. In this case, the Prandtl number is given by $P \cong (2/3)\,(T_0/T_e)^{1/2}\,(m_e/m_0)^{1/2}\,(Q_{ee}/Q_{00})$, where Q_{ee} and Q_{00} are the electron and atom transport cross sections, respectively. Since $T_0 \leqq T_e$, $m_e \ll m_0$, and in many cases $Q_{ee} \ll Q_{00}$, it is commonly $P \ll 1$.

First, examples of outflows are shown for the limiting (hypothetical) Prandtl number $P = 0$, which implies that heat conduction is the dominant energy transport mechanism. For $P = 0$, Eq. (19) can be integrated analytically to give $\varphi(\theta) = \cos 2m\theta > 0$ for symmetrical flows, $\varphi'(0) = 0$. The dependence of $g(\theta)$ on H is shown in Figs. 2—4 for $\gamma = 5/3$, $m = 1/4$, $R = 10^3$, $M = 10.30$ ($E = 71.11$, $s = 0.4$). In Fig. 4, only the maxima, minima, and the zero of $g(\theta)$ are indicated in the cases $H = 13.5$, 13.6. In Fig. 2, beyond a certain H, $g(\theta)$ becomes oscillatory with an increasing number of minima as H increases. Fig. 3 demonstrates how a new minimum enters as H increases. Since $g'' > 0$ (for $R > 0$) when $g = 0$, one cannot have a local minimum at $g = 0$. For the value of $H = 13.20$ (Fig. 3), $g(\theta)$ has three minima and is cutting the θ axis at θ_0, at about the closest point possible for three minima solutions $g(\theta)$ (here $\theta_0 \cong 1.82$). Increasing H moves θ_0 very rapidly towards the point where $\varphi(\theta) = 0$ ($\theta = \pi$ in this case) and a new minimum forms. As H continues to increase

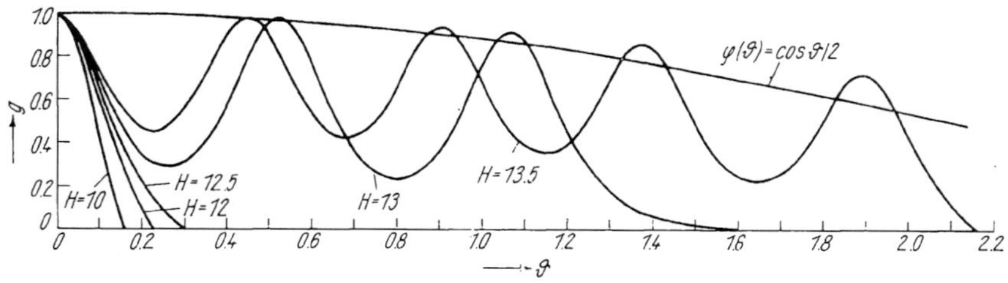

Fig. 2. $g(\theta)$ for various H, $\varphi'(0) = 0$, $\gamma = 5/3$, $m = 1/4$, $R = 1000$, $P = 0$, $M = 10.30$ ($E = 71.11$)

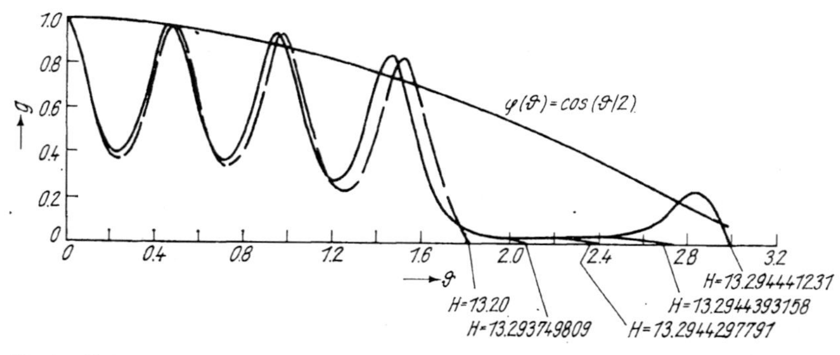

Fig. 3. $g(\theta)$ for various H, $\varphi'(0) = 0$, $\gamma = 5/3$, $m = 1/4$, $R = 1000$, $P = 0$, $M = 10.30$ ($E = 71.11$)

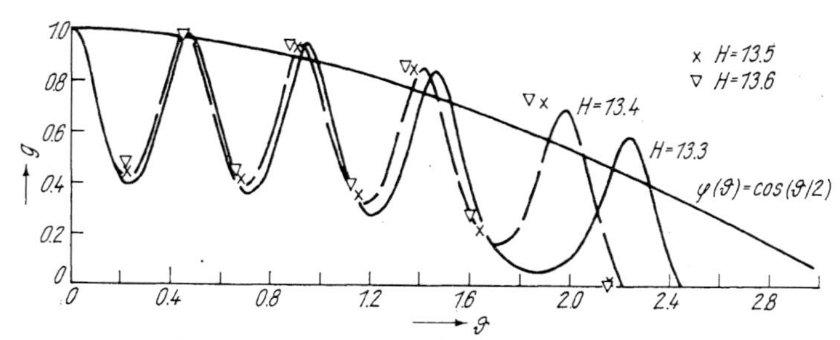

Fig. 4. $g(\theta)$ for various H, $\varphi'(0) = 0$, $\gamma = 5/3$, $m = 1/4$, $R = 1000$, $P = 0$, $M = 10.30$ ($E = 71.11$). For $H = 13.5$ and 13.6, only maxima, minima, and the zero of $g(\theta)$ are indicated

θ_0 decreases as can be seen in Fig. 4 up to about $\theta_0 = 2.15$ for $H \simeq 13.6$, and the process repeats itself. The number of minima is bounded since for $H^2 > (m+1)^2 + mR + (m+1)(7-m)(1-s)/3s$, $g(\theta)$ does not cut the θ axis for $\varphi(\theta) > 0$.

In the case of inflows with $P = 0$ and $\varphi'(0) = 0$, the dependence of $g(\theta)$ on the ECKERT number $E \equiv \equiv (\gamma - 1) M^2 = 3[(\gamma - 1)/\gamma(7-m)] Rs$ is shown in Fig. 5 [$\gamma = 5/3$, $m = -3/4$, $H = 0$, $R = -2000/3$] and Fig. 6 [$\gamma = 5/3$, $m = 1/2$, $R = -1000$]. In Fig. 6, only maxima, minima, and the last zero of $g(\theta)$ are shown in all cases $s \neq -0.0085$. It is seen that the inflows are mixed flows with inflow ($g > 0$, $R < 0$) and outflow ($g < 0$, $R < 0$) regions. The number of "oscillations" and the duct angle θ_0 increase with increasing $s < 0$ or $E > 0$.

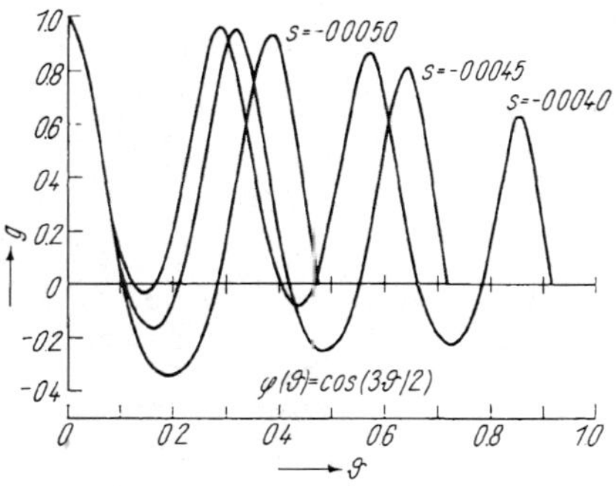

Fig. 5. $g(\theta)$ for various $E = 3[(\gamma - 1)/\gamma(7-m)] Rs$, $\varphi'(0) = 0$, $\gamma = 5/3$, $m = -3/4$, $R = -2000/3$, $P = 0$, $H = 0$

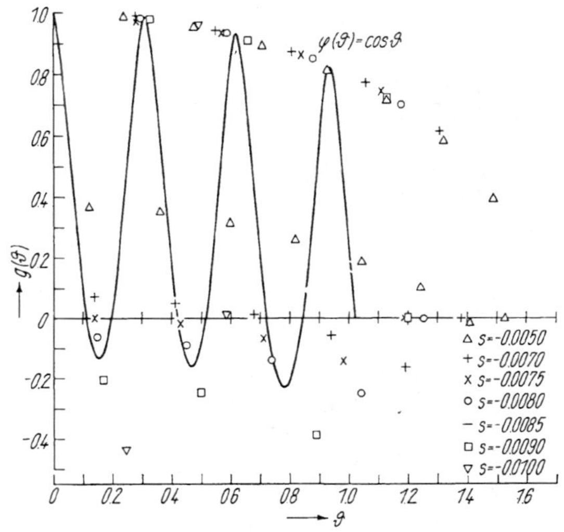

Fig. 6. $g(\theta)$ for various $E = 3[(\gamma - 1)/\gamma(7-m)] Rs$, $\varphi'(0) = 0$, $\gamma = 5/3$, $m = -1/2$, $R = -1000$, $P = 0$, $H = 0$. In all cases $s \neq -0.0085$, only maxima, minima, and the last zero of $g(\theta)$ are indicated

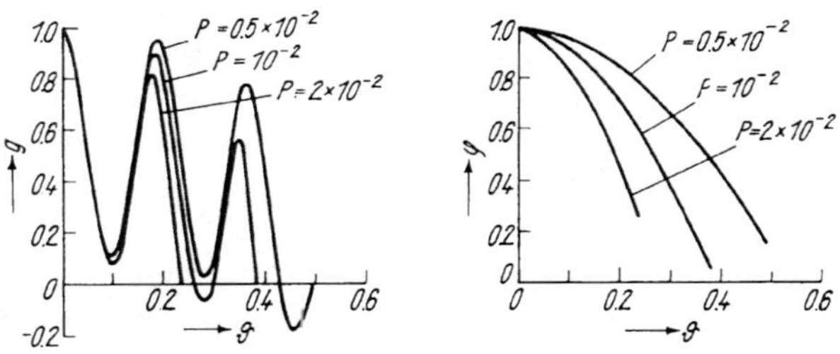

Fig. 7 $g(\theta)$ and $\varphi(\theta)$ for various P, $\varphi'(0) = 0$, $\gamma = 5/3$, $m = -1/2$, $R = -2000$, $H = 0$, $M = (2/3)^{1/2}$ ($E = 1$)

The dependence of inflow solutions $g(\theta)$ and $\varphi(\theta)$ on the PRANDTL number P is shown in Fig. 7 for $\gamma = 5/3$, $m = -1/2$, $H = 0$, $R = -2000$, $M = (2/3)^{1/2}$ [$E = 1$, $\varphi'(0) = 0$]. It is seen that the number of mixed flow regions $g(\theta) \lessgtr 0$ decreases and the $\varphi(\theta)$ profiles become steeper for increasing P.

Figure 8 compares symmetrical [$\varphi'(0) = 0$, $\varphi_+ = \varphi_-$] and asymmetrical [$\varphi'(0) = 10$, $\varphi_+ \neq \varphi_-$] solutions of $g(\theta)$ and $\varphi(\theta)$ for $\gamma = 5/3$, $m = 1/4$, $H = 5$, $R = 500$, $P = 10^{-5}$, $M = (2/3)^{1/2}$ ($E = 1$). As expected, the highly asymmetrical $\varphi(\theta)$ distribution gives rise to only a slight asymmetry in the $g(\theta)$ profile since P is small.

In the limit of large REYNOLDS numbers R, the solutions $g(x)$ and $\varphi(x)$, where $x = \theta/\theta_0$, are identical for different θ_0 and R as long as the combination $\alpha = \theta^2 R$ remains the same [cf. below]. Solutions $g(x)$ and $\varphi(x)$ of outflows ($m > 1$) and inflows ($m < -1$) are shown in Fig. 9 for $\gamma = 5/3$, $P = 10^{-1}$, $E = 1$, $H = 0$ and 1) $m = 4$, $\alpha = 0.412$ and 2) $m = -2$, $\alpha = -1.103$, respectively. Outflows and mixed flows are shown in Fig. 10 for $\gamma = 5/3$, $E = 1$, $H = 0$ and various m, $\alpha > 0$, and P in the region $-1 < m < 0$. The dependence of inflows and mixed flows $g(x)$ and corresponding $\varphi(x)$ on various $\alpha < 0$ is shown for $\gamma = 5/3$, $E = 1$, $H = 0$, $m = -3/5$ and different P, in Fig. 11. An example of outflow solutions $g(x)$ and $\varphi(x)$ with $\hat{x} \neq 0$ is shown in Fig. 12 for $\gamma = 5/3$, $m = -0.288$, $\alpha = 90.746$, $P = 10^{-2}$, $E = 0.961$, and $H = 0$.

These illustrations demonstrate the great variety of outflow and inflow solutions which can be expected to exist in magnetogasdynamic diffuser flows, presumed that these steady-state solutions are stable against arbitrary perturbations. The oscillatory and mixed flow solutions, which occur at relatively large duct angles, are particularly remarkable.

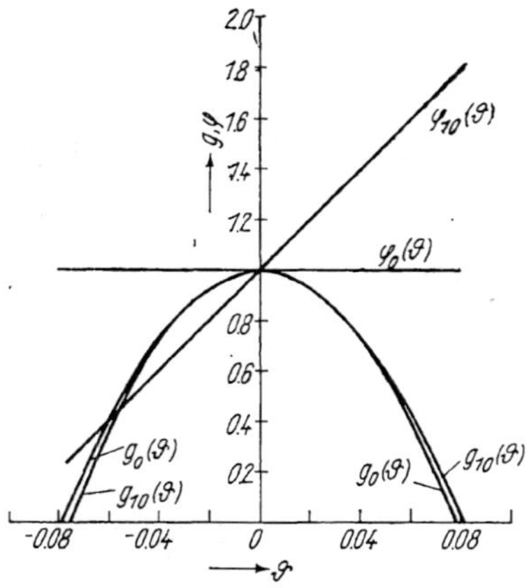

Fig. 8. $g(\theta)$ and $\varphi(\theta)$ for $\varphi'(0) = 0$ and $\varphi'(0) = 10$, $\gamma = 5/3$, $m = 1/4$, $R = 500$, $P = 10^{-5}$, $H = 5$, $M = (2/3)^{1/2}$ $(E = 1)$

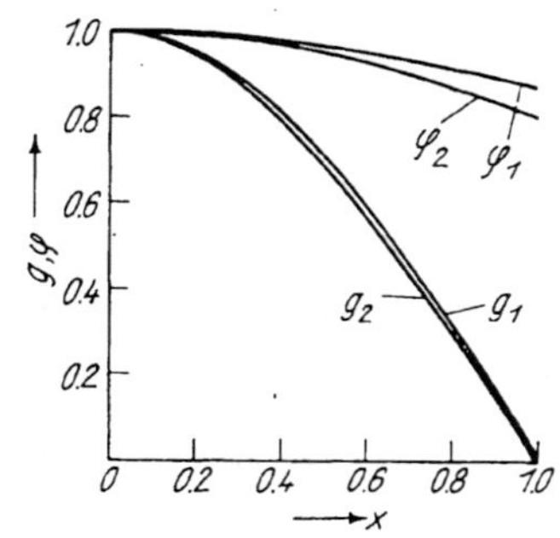

Fig. 9. $g(x)$ and $\varphi(x)$ for $\varphi'(0)=0$; $\gamma = 5/3$, $P = 10^{-1}$, $E = 1$, $H = 0$, and: 1) $m = 4$, $\alpha = 0.412$; 2) $m = -2$, $\alpha = -1.103$

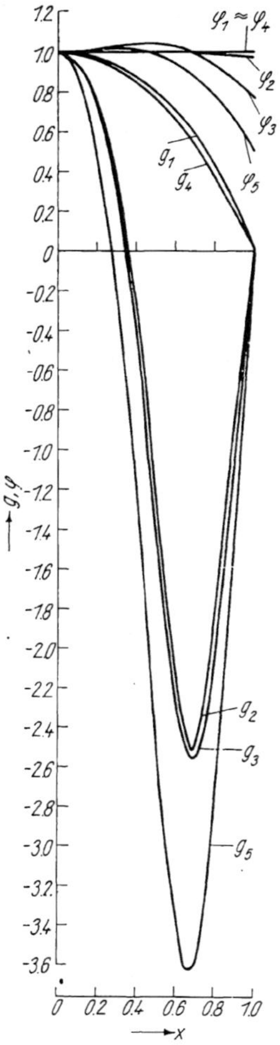

Fig. 10. $g(x)$ and $\varphi(x)$ for $\varphi'(0) = 0$, $\gamma = 5/3$, $E = 1$, $H = 0$, and:
1) $m = -1/5$, $\alpha = 12.070$, $P = 10^{-3}$;
2) $m = -1/5$, $\alpha = 99.268$, $P = 10^{-3}$;
3) $m = -1/5$, $\alpha = 95.709$, $P = 10^{-2}$;
4) $m = -1/10$, $\alpha = 7.005$, $P = 10^{-3}$;
5) $m = -1/10$, $\alpha = 94.640$, $P = 10^{-2}$

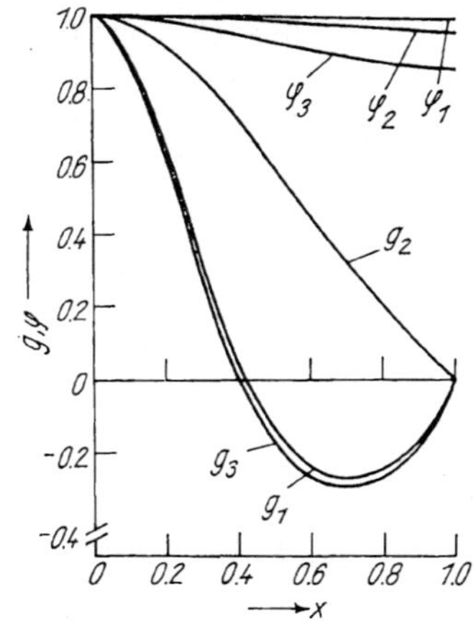

Fig. 11. $g(x)$ and $\varphi(x)$ for $\varphi'(0) = 0$, $\gamma = 5/3$, $m = -3/5$, $E = 1$, $H = 0$, and: 1) $\alpha = -52.963$, $P = 10^{-3}$; 2) $\alpha = -9.227$, $P = 10^{-2}$; 3) $\alpha = -57.500$, $P = 10^{-2}$

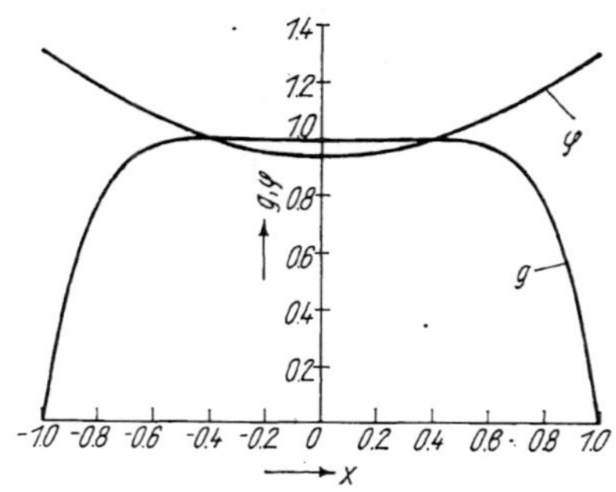

Fig. 12. $g(x)$ and $\varphi(x)$ for $\hat{x} = \hat{\theta}/\theta_0 \neq 0$, $\varphi'(0) = 0$ $\gamma = 5/3$, $m = -0.288$, $\alpha = 90.746$, $P = 10^{-2}$, $E = 0.961$, $H = 0$

Large REYNOLDS Numbers

Since it is clear that the larger $|R|$ is the smaller the possible θ_0 may be, let the independent variable be rescaled. The boundary-value problem becomes then:

$$g'' + \theta_0^2\,[(m+1)^2 - H^2]\,g + m\alpha g^2(1 - s + sg)\,\varphi^{-1} + \frac{\theta_0^2}{3}(m+1)(7-m)\frac{(1-s)}{s} = 0\,, \tag{31}$$

and

$$\varphi'' + (2m\theta_0)^2\,\varphi = -\,PE\left\{(g')^2 + \theta_0^2\left[\frac{4}{3}(m^2+m+1) + H^2\right]g^2 + \right.$$

$$\left. + \frac{\alpha(\gamma+1)}{\gamma E}\left[m - \frac{\gamma-1}{\gamma+1}\right](1 - s + sg)\,g\right\}, \tag{32}$$

where

$$x = \theta/\theta_0\,, \qquad (' \equiv \mathrm{d}/\mathrm{d}x)\,, \qquad g(\pm 1) = 0\,, \qquad \varphi(\pm 1) = \varphi_\pm\,, \qquad g(\hat{x}) = \varphi(\hat{x}) = 1\,, \quad (33)\,(34)$$

and

$$\alpha = \theta_0^2 R\,. \tag{35}$$

For large $|R|$, and for $|m| \ll |R|$, $H^2 \ll |R|$, Eqs. (31)—(32) may be approximated by

$$g'' + m\alpha g^2\varphi^{-1} + (m+1)\,\alpha\Gamma^{-1} = 0\,, \tag{36}$$

$$\varphi'' + (2m\theta_0)^2\,\varphi = -\,PE\left\{(g')^2 + \alpha\left[m\frac{\gamma+1}{\gamma-1} - 1\right]\Gamma^{-1}g\right\}. \tag{37}$$

If $\hat{x}$ is defined at a maximum of g, then $g''(\hat{x}) \leqq 0$, and by Eqs. (34) and (36)

$$[m + (m+1)\,\Gamma^{-1}]\alpha \geqq 0\,. \tag{38}$$

Hence,

$$m \geqq -\frac{1}{1+\Gamma} \quad \text{for} \quad \alpha > 0\,, \qquad m \leqq -\frac{1}{1+\Gamma} \quad \text{for} \quad \alpha < 0\,. \qquad (39)\ (40)$$

1. Monotone flows

If $m \geqq 0, \alpha > 0$ or $m < -1, \alpha < 0$ then

$$g'' = -\,m\alpha g^2 \varphi^{-1} - (m+1)\,\alpha\Gamma^{-1} \leqq -\,(m+1)\,\alpha\Gamma^{-1}\,, \tag{41}$$

and g is monotone decreasing for $x > \hat{x}$. Accordingly,

$$g \leqq -\frac{1}{2}(m+1)\,\alpha\Gamma^{-1}\,(x-\hat{x})^2 + 1\,. \tag{42}$$

By (33),

$$\frac{1}{2}(m+1)\,\alpha\Gamma^{-1}(\hat{x} \pm 1)^2 \leqq 1\,. \tag{43}$$

The least stringent condition occurs for $\hat{x} = 0$, which yields

$$(m+1)\,\alpha < 2\Gamma\,, \quad \text{or} \quad \alpha < \frac{2\Gamma}{m+1} \quad \text{for} \quad \alpha > 0\,, \qquad m > 0\,, \tag{44}$$

and

$$-\alpha < \frac{2\Gamma}{-\,(m+1)} \quad \text{for} \quad \alpha < 0\,, \qquad m < -1\,. \tag{45}$$

Additional restrictions are obtained from Eq. (31). If $\alpha > 0$ and $m \geqq \frac{\gamma-1}{\gamma+1}$ or $\alpha < 0$ and $m < -1$, then

$$\varphi'' < -\,PE\,(g')^2 \leqq -\,PE\,[(m+1)\,\alpha\Gamma^{-1}]^2\,(x-\hat{x})^2\,, \tag{46}$$

and

$$\varphi \leqq 1 + \varphi'(\hat{x})\,(x-\hat{x}) - PE(m+1)^2\,\alpha^2\Gamma^{-2}\,(x-\hat{x})^4/12\,. \tag{47}$$

Since $\varphi(\pm 1) \geqq 0$, Eq. (47) gives

$$P < \frac{12\gamma^2 E}{\alpha^2(m+1)^2\,(\gamma-1)^2}\,. \tag{48}$$

If $\alpha > 0$ and $0 < m < \frac{\gamma-1}{\gamma+1}$ then $\varphi'' < -\,PE(g')^2 + PE\alpha\left[1 - m\frac{\gamma+1}{\gamma-1}\right]\Gamma^{-1}g$. In this case,

$$P < 24\gamma^2 E\,\{\alpha^2\,[2(\gamma-1)^2\,(m+1)^2 - [12E(\gamma/\alpha) - (m+1)\,(\gamma-1)]\,[(\gamma-1) - m(\gamma+1)]]\}^{-1}\,. \tag{49}$$

2. Mixed flows

Solutions may show both inflow and outflow regions only if $0 > m > -1$, in particular only if

$$\frac{-1}{1+\Gamma_+} \leqq m < 0 \quad \text{for} \quad \alpha > 0\,, \tag{50}$$

or

$$-1 < m \leqq \frac{-1}{1+\Gamma_-} \quad \text{for} \quad \alpha < 0\,. \tag{51}$$

Hence,

$$\frac{-1}{1+\Gamma_+} < m \leqq \frac{-1}{1+\Gamma_-} \tag{52}$$

if a mixed flow exists. Here Γ_+ is computed with E defined at a positive maximum of u, $(R > 0)$, whereas for Γ_-, E is defined at a negative minimum of u, $(R < 0)$.

If a mixed flow exists, define $\hat{x}$ at the positive maximum of u, $(\alpha > 0)$. From Eq. (36) one obtains

$$g'' \geqq -\,(m+1)\,\alpha\Gamma^{-1}\,, \quad \text{and} \quad g \geqq 1 - \frac{1}{2}(m+1)\,\alpha\Gamma^{-1}\,(x-\hat{x})^2\,, \qquad (53)\ (54)$$

from which one concludes as before that $(m+1)\,\alpha\Gamma^{-1} \geqq 2$, or

$$\alpha \geqq \frac{2\Gamma}{m+1}\,(>0)\,. \tag{55}$$

This is also necessary for any outflow region as well, whenever $m < 0$. The condition in Eq. (48) now simplifies to

$$PE < \frac{1+m}{1-m} \cdot \frac{\gamma-1}{\gamma-(1+m)/(1-m)}\,(<1)\,. \tag{56}$$

Table I. (Plus sign for values at a local positive maximum, and minus sign for a local negative minimum)

m	Type of Flow	Restriction on Angle	Restriction on m	Other Restrictions
$m \geqq 0$	Pure outflow g convex	$\theta_0^2 < 2\Gamma_+ R_+^{-1}$	$m \leqq 2\Gamma_+ R_+^{-1}\theta_0^{-2} - 1$	
$m \geqq \frac{\gamma - 1}{\gamma + 1}$	φ convex			$P < \frac{12\Gamma_+^2}{R_+^2\theta_0^4(m+1)^2}$
$0 \leqq m < \frac{\gamma - 1}{\gamma + 1}$				Equation (49)
$m \leqq -1$	Pure inflow g convex φ convex		$m \geqq -2\Gamma_-(-R_-)^{-1}\theta_0^{-2} - 1$	$P < \frac{12\Gamma_-^2}{R_-^2\theta_0^4(m+1)^2}$
$-1 < m < 0$	Outflow or outflow region		$\frac{-1}{1 + \Gamma_+} \leqq m$	
		$\theta_0^2 > 2\Gamma_+ R_+^{-1}$	$2\Gamma_+ R_+^{-1}\theta_0^{-2} - 1 \leqq m$	
	Inflow or inflow region		$m \leqq \frac{-1}{1 + \Gamma_-}$	$P < \frac{\gamma}{2}\left[1 + \frac{1 - m}{-m}\left(\frac{\gamma - 1}{2}\right)\right]^{-1} \leqq \frac{1}{2}$
	Mixed flows	$\theta_0^2 > 2\Gamma_+ R_+^{-1}$	$\frac{-1}{1 + \Gamma_+} \leqq m \leqq \frac{-1}{1 + \Gamma_-}$	$P < \frac{\gamma}{2}\left[1 + \frac{1 - m}{-m}\left(\frac{\gamma - 1}{2}\right)\right]^{-1} \leqq \frac{1}{2 + E_+}$
			$2\Gamma_+ R_+^{-1}\theta_0^{-2} - 1 \leqq m$	$PE_+ < \frac{1 + m}{1 - m} \cdot \frac{\gamma - 1}{\gamma - (1 + m)/(1 - m)} < 1$

Theorem 1 and Corollary 2 are severe restrictions when $m > 1$ or $m < -1$ since Eq. (21) implies in these cases that $|R| < 4/3\,|m|$. As it is not physically feasible that m be very large, then $|R|$ must not be large, certainly much less than the order of 10^6 as typical of plasma flows. This means that if there is a backflow region when $m > 1$, or an outflow region when $m < -1$, the maximum velocity there is very small relative to velocities typical of the flow. One can conclude that most likely, $m \geqq 0$ will represent only pure outflows, and $m \leqq -1$ only pure inflows. In fact this is the case for large REYNOLDS numbers. The range of $-1 < m < 0$ allows mixed flows, as is shown numerically.

The results for large REYNOLDS numbers (which interest most) are summarized in Table I Note especially the upper bound on the possible angle for pure outflows when $m \geqq 0$, and the lower bound on the angle in the case of mixed flows, $-1 < m < 0$.

Further Remarks

The analysis presented indicates a manifold of flow types in magnetogasdynamic diffusers and confusers: pure outflows, pure inflows, mixed forward and back flows, and oscillatory outflows. These steady-state flow solutions are observable in an experiment only if they are stable against arbitrary perturbations. The mixed forward and back flow distributions and the associated temperature distributions are qualitatively similar to those obtained by analytical methods for the incompressible HAMEL-flow with heat transfer by MILLSAPS and POHLHAUSEN [4]. The finger-type outflows at large duct angles with the velocity amplitude oscillating between the limits $0 \leqq g(\theta) \leqq 1$ represent new flow types, which have apparently not yet been observed in experiments.

A comparison indicates that the pure outflow solutions can be modeled in good approximation by means of the polytropic approximation applied previously to compressible diffuser flows [7]. On the other hand, the polytropic approximation is not adequate for the description of inflows and the more complicated outflow types [7]. This shows the limitations of the polytropic approximation which is frequently used in (mathematical) gasdynamics.

The present theory for magnetogasdynamic flows between inclined walls leads to solutions which are unique and exhibit no homogeneous over-pressure fields. The lack of solutions with homogeneous over-pressure is also typical for the polytropic diffuser theory [7]. On the other hand, the classical HAMEL theory [2] for isothermal, incompressible fluids gives up to an infinite number of different solutions for one and the same set of values θ_0, R, and Q. The non-uniqueness of the HAMEL solutions is related to the existence of solutions with over-pressures in the incompressible case. These homogeneous over-pressures p_0 are chosen sufficiently large so that the variable pressure component $p(r, \theta)$, which is negative within certain regions of $|\theta| \leqq \theta_0$ for the more complicated flow types, is balanced to a total pressure $P(r, \theta) = p_0 + p(r, \theta) > 0$, $|\theta| \leqq \theta_0$ and $0 < r < \infty$.

Interest in this theoretical problem arose in connection with experiments on compressible plasma flows across inhomogeneous magnetic fields in diverging ducts [11, 12]. In the limit of vanishing electrical conductivity, the theory presented describes gas flows in diffusers and confusers, a previously unsolved problem of ordinary gas dynamics.

Appendix: Theorems on the Possible Flows

The parameters H, P and γ do not depend on the particular value of $\hat{\theta}$ chosen, whereas M, E, R and therefore also s do depend on the choice of $\hat{\theta}$. M^2 and E are always positive, and $R \gtrless 0$ as $u_0 \gtrless 0$, i.e., according to whether at the reference point there is outflow or inflow. Thus $Rg > 0$ represents an outflow region, and $Rg < 0$ represents an inflow region. The freedom to select $\hat{\theta}$ will prove useful in the following theorems.

Theorem 1: *A necessary condition for pure inflow, or an inflow region in a mixed flow is*

$$-mR \leqq \frac{4}{3}(m^2-1) + (m+1)\frac{R}{\gamma M^2} - H^2\,, \tag{A1}$$

where $\hat{\theta}$ is a point where $u(r_0, \theta)$ achieves a local minimum in that region ($R < 0$).

Proof: By the choice of $\hat{\theta}$, $g(\hat{\theta}) = 1$, $g'(\hat{\theta}) = 0$, and $g''(\hat{\theta}) \leqq 0$, since $g(\theta)$ achieves a maximum at $\theta = \hat{\theta}$. By Eqs. (18), (25), and (23) we have $g''(\hat{\theta}) = -mR + H^2 - (m+1)^2 - \frac{1}{3}(m+1)(7-m)\frac{1-s}{s} = -mR + H^2 + \frac{4}{3}(1-m^2) - (m+1)\frac{R}{\gamma M^2}$. Hence, Eq. (A1) holds.

Corollary 1: *If $0 \leqq m \leqq 1$, only pure outflow can exist.*

Proof: If an inflow region exists, then the left hand side of Eq. (A1) is positive and the right hand side negative. Thus, the assumption of an inflow gives a contradiction.

The case $0 < m < 1$ is of special interest. In the polytropic case it is $p = \alpha\varrho^k$, $m = (k-1)/(k+1)$, $1 < k < \infty$, and hence $0 < m < 1$. Thus, polytropicity implies pure outflows only [7].

Corollary 2: *A necessary condition for pure outflow (or an outflow region) is Eq. (A1), where $\hat{\theta}$ is a point where $u(r_0, \theta)$ achieves a local maximum in that region ($R > 0$).*

Proof: Identical to proof of Theorem 1.

Corollary 3: *For $m = -1$ only pure inflow can exist.*

Proof: From Eq. (A1) one gets $R \leqq -H^2 \leqq 0$, but by Corollary 2, one would have $R > 0$ if an outflow region existed.

Corollary 4: *If $-1 < m < 0$, necessary conditions for pure inflow or an inflow region are*

$$mR > \frac{4}{3}(1-m^2)\,, \qquad (m+1)\frac{(-R)}{\gamma M^2} < mR - \frac{4}{3}(1-m^2)\,, \tag{A2) (A3}$$

and

$$H^2 < mR - \frac{4}{3}(1-m^2) - (m+1)\frac{(-R)}{\gamma M^2}\,, \tag{A4}$$

$\hat{\theta}$ taken at a local minimum of $u(r_0, \theta)$ in the region where $R < 0$.

Proof: Immediate from Theorem 1.

Corollary 5: *If $-1 < m < 0$, a necessary condition for pure outflow or an outflow region is*

$$\frac{-1}{1+\gamma M^2} < m\ (< 0)\,, \tag{A5}$$

where $\hat{\theta}$ is taken at any local positive maximum of $u(r_0, \theta)$ where $R > 0$. Hence, a necessary condition for pure inflow or an inflow region is

$$(-1 <)\ m < \frac{-1}{1+\gamma M^2}\,, \tag{A6}$$

where $\hat{\theta}$ is taken at any local negative minimum of $u(r_0, \theta)$ where $R < 0$.

Proof: By Theorem 1 and Corollary 2, Eq. (A1) holds. Since $m^2 < 1$, it is $-mR < (m+1)R/\gamma M^2$, and Eqs. (A5)–(A6) follow.

So far only Eq. (18) has been used. By means of Eqs. (18)–(20) one can get restrictions on P, too. Equations (18)–(20) give

$$\begin{aligned}(\varphi g' - g\varphi')' = \varphi g'' - g\varphi'' = {} & [-S - (1-m)(1+3m)g]\,\varphi + H^2 g\varphi + PEg'^2 g + \\ & + PE\left[\frac{4}{3}(1+m+m^2) + H^2\right]g^3 + \\ & + (1-s+sg)\,g^2\left\{\frac{2}{\gamma}PmR - mR - (1-m)\,RP\Gamma^{-1}E\right\},\end{aligned} \tag{A7}$$

where

$$S \equiv \frac{1}{3}(1+m)(7-m)(1-s)/s\,. \tag{A8}$$

Integrating Eq. (A7) in a region (θ_1, θ_2) where g is of fixed sign and $g(\theta_1) = g(\theta_2) = 0$, one finds

$$\varphi(\theta_2)\, g'(\theta_2) - \varphi(\theta_1)\, g'(\theta_1) = \int_{\theta_1}^{\theta_2} [-S - (1-m)(1+3m)\, g]\, \varphi\, d\theta + \\ + \int_{\theta_1}^{\theta_2} \left\{ H^2\varphi + PE\left[(g')^2 + \left[\frac{4}{3}(1+m+m^2) + H^2\right] g^2 \right]\right\} g\, d\theta + \\ + R\left[\left(\frac{2}{\gamma}P - 1\right)m - (1-m)\, P\Gamma^{-1}E\right] \int_{\theta_1}^{\theta_2} (1 - s + sg)\, g^2\, d\theta\,. \tag{A9}$$

The left hand side of Eq. (A9) is $\leqq 0$ if $g > 0$ in the region, and is $\geqq 0$ if $g < 0$. Notice that in Eq. (A9) the second integral has the same sign as g, and the third integral is always positive, by Eqs. (26) and (28). If,

$$-S - (1-m)(1+3m)\, g \geqq 0 \tag{A10}$$

for $g > 0$ in the region then, since the left side of Eq. (A9) is not positive, the coefficient of the third integral must be negative, i.e.,

$$R\left[\left(\frac{2}{\gamma}P - 1\right)m - (1-m)\, P\Gamma^{-1}E\right] < 0\,. \tag{A11}$$

Also, if

$$-S - (1-m)(1+3m)\, g \leqq 0 \tag{A12}$$

when $g < 0$ in the region, then the corresponding necessary condition is

$$R\left[\left(\frac{2}{\gamma}P - 1\right)m - (1-m)\, P\Gamma^{-1}E\right] > 0\,. \tag{A13}$$

From Eqs. (A10) and (A13) the following theorems are obtained.

Theorem 2: *If $m > 1$, a necessary condition for the existence of pure inflow or mixed (in and out) flows is*

$$P > \frac{\gamma}{2}\left[1 + \left(\frac{m-1}{m}\right)\left(\frac{\gamma-1}{2}\right)\right]^{-1} > \frac{\gamma}{\gamma+1} \geqq \frac{1}{2}\,. \tag{A14}$$

Proof: By Eq. (25), it is $Rs\,(7-m) > 0$. If $1 < m \leqq 7$, then $-S \geqq 0$ if and only if $s < 0$, and if $7 < m$, then $-S \geqq 0$ if $s > 0$. Each of the four cases $R \gtrless 0$, $m \gtrless 7$ should be checked separately, but all yield the same result. As an example, take the case $1 < m < 7$ and $R > 0$. This implies $s > 0$, and $-S < 0$. The condition $-S < 0$ is necessary for Eq. (A12) to hold and is also sufficient for Eq. (A10) not to hold. If one looks at a region for which $g < 0$, then Eq. (A12) holds. Since $R > 0$, there must also be an outflow region ($g > 0$). Hence, this case means a mixed flow. Eq. (A13) now follows, which gives

$$\left(\frac{2}{\gamma}P - 1\right)m - (1-m)\, P\frac{\gamma - 1}{\gamma} > 0\,, \tag{A15}$$

and Eq. (A14).

Theorem 3: *If $-1 \leqq m < 0$, necessary conditions for the existence of a pure inflow or a mixed flow are,*

$$(0 \leqq)\, P < \frac{\gamma}{2}\,, \tag{A16}$$

$$P < \frac{-m}{1-m}\left\{1 - \frac{1+m}{\gamma(1-m)}\right\}^{-1} = \frac{\gamma}{2}\left\{1 + \frac{1-m}{-m}\left(\frac{\gamma-1}{2}\right)\right\}^{-1} \leqq \frac{1}{2}\,. \tag{A17}$$

Proof: For $-1 \leqq m < 0$, and inflow, one has $g > 0$, $R < 0$, $s < 0$ and $-S \geqq 0$. For $(1+3m) \leqq 0$, it is $-S - (1-m)(1+3m)\, g \geqq 0$. For $1 + 3m > 0$, Eq. (30) gives

$$\frac{1}{3}(1+m)(7-m)\left(1 - \frac{1}{s}\right) - (1-m)(1+3m)\, g > \frac{1}{3}(1+m)(7-m)\left(1 - \frac{1}{s}\right) - \\ - (1-m)(1+3m)\left(1 - \frac{1}{s}\right) = \frac{4}{3}(1+2m^2)\left(1 - \frac{1}{s}\right) > 0\,. \tag{A18}$$

Eq. (A10), and therefore (A11) hold, from which Eq. (A14) again follows. Since m is negative, Eq. (A16) must hold, and Eq. (A17) is equivalent to Eq. (A14).

By the same argument one proves also

Theorem 4: *If* $m < -1$, *a necessary condition for pure outflow or mixed flow is Eq.* (A14), *i.e.*,

$$P > \frac{\gamma}{2}\left[1 + \frac{1-m}{-m}\left(\frac{\gamma-1}{2}\right)\right]^{-1} \geqq \frac{1}{2}. \tag{A19}$$

Since the combination PE also appears in Eqs. (19)—(20), it is desirable to get a restriction on it too.

Theorem 5: *If* $-1 < m < 0$, *a necessary condition for a mixed flow to exist is*

$$PE < \frac{1+m}{1-m} \cdot \frac{\gamma - 1}{\gamma - (1+m)/(1-m)} \cdot \frac{-mR}{-mR + H^2 + \frac{4}{3}(1-m^2)} < 1, \tag{A20}$$

where $\hat{\theta}$ *is any point where* $u(r_0, \theta)$ *achieves a positive maximum* ($R > 0$).

Proof: If a mixed flow exists then, by Corollary 2, Eq. (A1) holds with $\hat{\theta}$ defined as above. Multiplying by E gives

$$\left[\frac{4}{3}(1-m^2) + H^2 - mR\right] E \leqq (m+1)\frac{RE}{\gamma M^2} = (m+1) R \frac{\gamma-1}{\gamma}, \tag{A21}$$

and, since $-mR > 0$,

$$E \leqq \frac{(m+1) R(\gamma-1)/\gamma}{\frac{4}{3}(1-m^2) + H^2 - mR}. \tag{A22}$$

~~By Theorem 3, Eq. (A17) holds. Multiplying~~ the above inequality by the inequality in Eq. (A17) and rearranging terms yields Eq. (A20), since each of the three terms is less than one.

Similarly there follows

Theorem 6: *The condition*

$$PE > \frac{m+1}{m-1} \cdot \frac{\gamma - 1}{\gamma + (m+1)/(m-1)} \cdot \frac{-mR}{\frac{4}{3}(m^2-1) - (-mR) - H^2}, \tag{A23}$$

is necessary for inflow when $m > 1$ *and* $\hat{\theta}$ *is taken at a local negative minimum of* $u(r_0, \theta)$, *and is necessary for outflow when* $m < -1$ *and* $\hat{\theta}$ *is taken at a local positive maximum of* $u(r_0, \theta)$.

References

1 LORD RAYLEIGH, Phil. Mag. **34**, 94 (1917).
VON KARMAN, TH., ZAMM **1**, 233 (1921).
GUDERLEY, G., Z. Luftfahrtforschung **19**, 302 (1942).
SAKURAI, A., J. Phys. Soc. Japan **8**, 662 (1953); **9**, 256 (1954).
VON WEIZSAECKER, C. F., Z. Naturforsch. **9 a**, 269 (1954).
HAEFELE, W., Z. Naturforsch. **10a**, 1006 (1955).
BLUMAN, G. W., and COLE, J. D., J. Math. Mech. 18, 1025 (1969).
NARIBOLI, G. A., Appl. Sci. Res. 22, 449 (1970).
MILLSAPS, K., and NYDAHL, J. E., ZAMM **53**, 241 (1973).
2 HAMEL, G., Jahresbericht. Deutsch. Math. Vereinigung, 34 (1916).
3 LANDAU, L. D., Dokl. Akad. Nauk SSSR **43**, 299 (1944).
4 MILLSAPS, K., and POHLHAUSEN, K., J. Aeronaut. Sci. **20**, 187 (1953).
5 WILHELM, H. E., Phys. Fluids **17**, 360 (1974).
6 WILHELM, H. E., J. Math. Phys. **14**, 1430 (1973).
7 LIRON, N., and WILHELM, H. E., ZAMM **56**, 361 (1976).
8 SEDOV, L. I., Similarity and Dimensional Methods in Mechanics, Academic Press, New York 1969.
9 STANYUKOVICH, K. P., Unsteady Motion of Continuous Media, Academic Press,New York 1969.
10 WILHELM, H. E., Phys. Fluids **17**, 1841 (1974).
11 ZAUDERER, B., Phys. Fluids **11**, 2577 (1968).
12 ZAUDERER, B., and TATE, E., AIAA J. 9, 1136 (1971).

Eingereicht am 22. 12. 74, revidierte Fassung am 14. 1. 75.

Anschriften: Dr. N. LIRON, Department of Applied Mathematics, Weizmann Institute of Science, Rehovot, Israel
Prof. H. E. WILHELM, Department of Electrical Engineering, Colorado State University, Fort Collins, Colorado, 80523, USA

ZAMM 56, 491 – 497 (1976)

H. Wilke

Zur iterativen Berechnung nicht-Newtonscher Strömungen in Rohren mit Kreisringquerschnitt

Das Problem der Strömung eines nicht-Newtonschen Mediums durch ein Rohr mit Kreisringquerschnitt wird mit Hilfe einer numerischen Methode untersucht. Als Voraussetzung wird lediglich die Monotonie und Lipschitz-Stetigkeit der Materialfunktion gefordert. Unter diesen Bedingungen ist die Anwendung eines Projektions-Iterationsverfahrens zur Lösung des stationären und instationären Problems möglich. Anhand konkreter Berechnungsfälle für das Materialgesetz von Williamson und für das Potenzgesetz können qualitative und quantitative Angaben über das Strömungsverhalten gemacht werden.

The present paper is concerned with the numerical solution of the equation of motion for the axial flow of an incompressible, non-Newtonian fluid in a long cylindrical annulus. It is only supposed that the material law of the fluid is Lipschitzian and monotone. Under these conditions the non-linear flow problem can be solved numerically by means of a projection-iteration method. The initial value problem as well as the steady-state problem are considered for constant and periodic pressure gradient. Numerical calculations are performed to find the velocity profiles, velocity gradients and volume rates of flow for pseudoplastic fluids of Williamson-type and power law-type.

С помощью численного метода исследуется задача о течении некоторой неньютоновской среды через трубу с кольцевым поперечным сечением. В качестве предположения требуется лишь монотонность и непрерывность по Липшицу соответствующей функции материала. При выполнении этих условий становится возможным применение некоторого проекционно-интерационного метода для решения стационарной и нестационарной проблемы. На основе конкретно вычисленных примеров для закона материала Виллиамсона и потенциального закона материала стало возможным получение количественных и качественных данных о поведении течения.

1. Einführung

Gegenwärtig gewinnt der Transport nicht-Newtonscher (rheologischer) Medien in Rohren immer mehr an Bedeutung für die chemische Technologie. Strömungen von Ölen, Polymerschmelzen- und Lösungen lassen sich nicht mit Hilfe herkömmlicher Methoden, die den linearen Newtonschen Schubspannungsansatz enthalten, berechnen. Es entstehen vielmehr nichtlineare partielle Differentialgleichungen, für die keine geschlossenen analytischen Lösungen angegeben werden können.

Nachstehend wird die Strömung eines rheologischen Mediums durch ein unendlich langes Rohr mit Kreisringquerschnitt untersucht. Diese sogenannte Ringspaltgeometrie findet man u. a. in der Verarbeitungstechnik bei der Herstellung von Plasterohren und in der Verfahrenstechnik bei Wärmeübertragern. Zur Lösung des Problems wird ein Projektions-Iterationsverfahren verwendet, dessen numerische Realisierung ausführlich beschrieben wird. Für zwei konkrete Materialfunktionen, das Gesetz von Williamson und das Potenzgesetz, werden Lösungen angegeben und diskutiert.

2. Mathematische Beschreibung des Problems

Es wird die laminare Strömung eines inkompressiblen Mediums durch ein Rohr mit Kreisringquerschnitt betrachtet. Der Radius des inneren Rohres sei R_i, der Radius des äußeren Rohres R_a. Die instationäre Bewegungsgleichung für eine beliebige Materialfunktion $\mu\left(\left|\frac{\partial u}{\partial r}\right|\right)$ hat dann folgendes Aussehen:

$$\frac{\partial u}{\partial t} - \frac{1}{\varrho r}\frac{\partial}{\partial r}\left(r\mu\left(\left|\frac{\partial u}{\partial r}\right|\right)\frac{\partial u}{\partial r}\right) + \frac{1}{\varrho}\frac{\partial p}{\partial z} = 0\,, \tag{2.1}$$

ϱ – Dichte, $\frac{\partial p}{\partial z}$ – Druckgradient, $t \in [0, T]$, $r \in [R_i, R_a]$.

Die Geschwindigkeiten müssen folgenden Rand- und Anfangsbedingungen genügen:

$$u(R_i, t) = u(R_a, t) = 0\,, \quad u(r, 0) = w(r)\,. \tag{2.2}$$

$w(r)$ – Anfangsgeschwindigkeitsverteilung.

Die folgenden Hilbert-Räume V und H werden benötigt:

$$V = \left\{u \,\middle|\, \int_{R_i}^{R_a} r \mid u'(r)|^2 \,\mathrm{d}r < \infty,\ u(R_i) = u(R_a) = 0\right\}$$

mit dem Skalarprodukt

$$((u, v)) = \int_{R_i}^{R_a} ru'(r)\, v'(r)\,\mathrm{d}r$$

bzw.

$$H = \left\{ u \,\middle|\, \int_{R_i}^{R_a} r \mid u(r) \mid^2 \mathrm{d}r < \infty \right\}$$

mit

$$(u, v) = \int_{R_i}^{R_a} ru(r)\, v(r)\, \mathrm{d}r\,.$$

Hier und im folgenden bezeichnet u' die Ableitung von $u \in V$ im Sinne der Distributionen.

Der Raum V liegt in H dicht. Daher kann H als Teilraum des zu V dualen Raums V^* aufgefaßt werden. Für den Wert eines linearen Funktionals $f \in V^*$ im Punkt $u \in V$ schreiben wir (f, u). Das kann nicht zu Mißverständnissen führen, da für $f \in H$ der Wert des Funktionals f im Punkt u mit dem Skalarprodukt von f und u in H übereinstimmt. Wählt man $Ju = -(ru')'/r$, so gilt die Beziehung $(Ju, v) = \big((u, v)\big)$, $u, v \in V$, d. h. J ist die sogenannte Dualitätsabbildung von V auf V^*.

Es sei ferner

$$Au = -\frac{1}{\varrho r}\big(r\mu(|u'|)\, u'\big)' + \frac{1}{\varrho}\frac{\partial p}{\partial z} \quad \text{für} \quad u \in V\,.$$

Unter Benutzung von A läßt sich die Aufgabe (2.1), (2.2) folgendermaßen als Evolutionsgleichung schreiben:

$$\frac{\partial u}{\partial t} + Au = 0\,, \quad u(0) = a\,, \quad u \in V\,. \tag{2.3}$$

Unter der Voraussetzung, daß der Operator A monoton und LIPSCHITZ-stetig ist, läßt sich (2.3) nach [1] mit Hilfe eines Projektions-Iterationsverfahrens lösen. Die geforderte Monotonie und LIPSCHITZstetigkeit von A ist erfüllt, wenn gilt:

$$\begin{aligned} &\mu(\xi)\,\xi - \mu(\eta)\,\eta \geqq m(\xi - \eta) \quad \text{für} \quad \xi \geqq \eta \geqq 0,\ m > 0\,, \\ &|\mu(\xi)\,\xi - \mu(\eta)\,\eta| \leqq M\,|\xi - \eta| \quad \text{für} \quad \xi, \eta \geqq 0\,. \end{aligned} \tag{2.4}$$

Die Konstanten M und m lassen sich somit als maximaler bzw. minimaler Anstieg der Materialfunktion $\mu(|u'|)\,u'$ interpretieren. Nach [1] existiert ein in V orthonormales vollständiges System von Funktionen $\{h_1, h_2, \ldots\}$ mit den Eigenschaften:

$$Jh_j = \lambda_j h_j,\ \lambda_j > 0,\ (h_j, h_k) = \delta_{j,k}\,. \tag{2.5}$$

Die Iterationsvorschrift für die n-te Näherung u_n hat nach [1] dann folgende Gestalt:

$$\left(\frac{\partial u_n(t)}{\partial t} + aJu_n(t), h_j\right) = \big(aJu_{n-1}(t) - Au_{n-1}(t), h_j\big), \quad u_n(0) = w\,,$$

für $j = 1, \ldots, N(n)$, $n = 1, 2, \ldots$, u_0 beliebig.

In dieser Gleichung ist a ein reeller Parameter, der der Nebenbedingung $a > \dfrac{M}{2\varrho}$ genügen muß.

Da A ein Potentialoperator ist, erhält man für

$$a = \frac{M + m}{2\varrho}$$

die bestmögliche Kontraktionskonstante (siehe [2])

$$q = \frac{M - m}{M + m}.$$

Mit der Koordinatendarstellung

$$u_n(t) = \sum_{k=1}^{N(n)} c_{n,k}(t)\, h_k, \qquad w = \sum_{k=1}^{\infty} b_k h_k \tag{2.6}$$

folgt dann unter Beachtung der Gln. (2.5)

$$\frac{\partial c_{n,j}(t)}{\partial t} + a\lambda_j c_{n,j}(t) = a\lambda_j c_{n-1,j}(t) - \big(Au_{n-1}(t), h_j\big), \qquad c_{n,j}(0) = b_j\,,$$

oder

$$c_{n,j}(t) = \exp(-a\lambda_j t)\left(b_j + \int_0^t \exp(a\lambda_j s)\,\big(a\lambda_j c_{n-1,j}(s) - (Au_{n-1}(s), h_j)\big)\,\mathrm{d}s\right). \tag{2.7}$$

Dabei ist

$$\big(Au_{n-1}(s), h_j\big) = \frac{1}{\varrho}\int_{R_i}^{R_a} r\left(\mu\left(\left|\sum_{k=1}^{N(n-1)} c_{n-1,k}(s)\, h_k'(r)\right|\right)\sum_{k=1}^{N(n-1)} c_{n-1,k}(s)\, h_k'(r)\, h_j'(r) + \frac{\partial p}{\partial z} h_j(r)\right)\mathrm{d}r$$

$j = 1, \ldots, N(n)$, $c_{n-1,j} = 0$ für $j > N(n-1)$.

Damit ist der Lösungsalgorithmus für das Anfangswertproblem (2.3) gegeben. Nach [1] konvergiert das Projektions-Iterationsverfahren aber auch für den Fall, daß eine periodische Lösung gesucht wird. Anstelle der Gln. (2.2) stehen dann die Bedingungen

$$u(R_i, t) = u(R_a, t) = 0\,, \quad u(r, 0) = u(r, T)\,, \tag{2.8}$$

und die Aufgabe (2.3) ist durch die folgende zu ersetzen:

$$\frac{\partial u}{\partial t} + Au = 0\,, \quad u(0) = u(T)\,. \tag{2.9}$$

Die den Gln. (2.7) entsprechende Iterationsvorschrift lautet in diesem Fall

$$c_{n,j}(t) = \exp(-a\lambda_j t)\left(\frac{1}{\exp(a\lambda_j T) - 1}\int\limits_0^T \exp(a\lambda_j s)\left(a\lambda_j c_{n-1,j}(s) - (Au_{n-1}(s), h_j)\right) ds + \right.$$

$$\left. + \int\limits_0^t \exp(a\lambda_j s)\left(a\lambda_j c_{n-1,j}(s) - (Au_{n-1}(s), h_j)\right) ds\right). \tag{2.10}$$

Die Bestimmung der Basisfunktionen h_j führt nach Gl. (2.5) auf das Eigenwertproblem

$$Jh_j(r) = -\frac{1}{r}\left(rh_j(r)'\right)' = \lambda_j h_j(r)\,. \tag{2.11}$$

Die allgemeine Lösung dieser BESSELschen Differentialgleichung ist

$$h_j(r) = C_j J_0\left(\sqrt{\lambda_j}\, r\right) + D_j N_0\left(\sqrt{\lambda_j}\, r\right), \tag{2.12}$$

J_0 — BESSELsche } Funktion
N_0 — NEUMANNsche } nullter Ordnung.

Die Randbedingungen (2.2) bzw. (2.8) liefern:

$$C_j J_0\left(\sqrt{\lambda_j}\, R_i\right) + D_j N_0\left(\sqrt{\lambda_j}\, R_i\right) = 0\,, \qquad C_j J_0\left(\sqrt{\lambda_j}\, R_a\right) + D_j N_0\left(\sqrt{\lambda_j}\, R_a\right) = 0\,. \tag{2.13}$$

Aus der Bedingung für die nichttriviale Lösbarkeit von (2.13) erhält man die Bestimmungsgleichung für die Eigenwerte λ_j:

$$J_0\left(\sqrt{\lambda_j}\, R_i\right) N_0\left(\sqrt{\lambda_j}\, R_a\right) - J_0\left(\sqrt{\lambda_j}\, R_a\right) N_0\left(\sqrt{\lambda_j}\, R_i\right) = 0\,. \tag{2.14}$$

Für die Koeffizienten wird gesetzt:

$$C_j = \frac{N_0\left(\sqrt{\lambda_j}\, R_i\right)}{W_j}\,, \qquad D_j = -\frac{J_0\left(\sqrt{\lambda_j}\, R_i\right)}{W_j}\,. \tag{2.15}$$

Der Normierungsfaktor ergibt sich aus Gl. (2.5) unter Beachtung von

$$|h_j| = \left(\sqrt{h_j, h_j}\right) = 1$$

zu

$$W_j^2 = \int\limits_{R_i}^{R_a} r\left(N_0(\sqrt{\lambda_j}\, R_i)\, J_0(\sqrt{\lambda_j}\, r) - J_0(\sqrt{\lambda_j}\, R_i)\, N_0(\sqrt{\lambda_j}\, r)\right)^2 dr\,. \tag{2.16}$$

Für eine konstante Materialfunktion, die den Forderungen (2.4) genügt, sind damit alle Voraussetzungen für eine numerische Realisierung des Projektions-Iterationsverfahrens gegeben.

3. Numerische Realisierung

Zur numerischen Realisierung des Projektions-Iterationsverfahrens werden die Intervalle $[0, T]$ und $[R_i, R_a]$ durch Einführung eines konstanten Zeit- bzw. Ortsschrittes diskretisiert. Eine periodische Erregung wird durch folgende Form des Druckgradienten simuliert:

$$\frac{\partial p}{\partial z} = P_0 + P_A \sin\frac{2\pi t}{T}\,. \tag{3.1}$$

Als Erregung könnte aber auch jede andere stetige [0, T]-periodische Funktion dienen. Insbesondere liefert der triviale Fall $P_A = 0$ bei Anwendung auf das Problem (2.8) sofort die stationäre Lösung des Problems (2.3).

Zur Erläuterung der Ausführung der in Gl. (2.7) bzw. (2.10) beschriebenen Iterationsvorschriften sollen die folgenden Bemerkungen dienen.

— Berechnung der Integrale:

Die Integration des ersten Summanden im Ortsintegral der Gl. (2.7) gelingt nur mittels einer numerischen Methode (es wurde die Trapezformel verwendet), während der zweite Summand exakt integriert werden kann.

Das Zeitintegral in Gl. (2.7) bzw. (2.10) hat die Form

$$\int_0^t g(s)\, f(s)\, \mathrm{d}s\,, \quad g(s) = \exp\,(a\lambda_j s)\,.$$

Es kann dargestellt werden als Summe der Integrale der einzelnen Zeitschritte $\Delta t = t_i - t_{i-1}$

$$\int_{t_{i-1}}^{t_i} g(s)\, f(s)\, \mathrm{d}s\,, \quad i = 1, 2, \ldots, I(t)\,.$$

Schreibt man dieses Integral näherungsweise als

$$\int_{t_{i-1}}^{t_i} g(s) \left(f(t_i)\, \frac{s - t_{i-1}}{\Delta t} + f(t_{i-1})\, \frac{t_i - s}{\Delta t} \right) \mathrm{d}s\,,$$

so können die beiden Summanden integriert werden.

— Die Materialfunktion

Der konkrete Zusammenhang $\mu(|u'|)$ wird mit Hilfe einer Funktionsprozedur verwirklicht. Als Parameter werden nur ein Feld von Konstanten sowie der gerade aktuelle Geschwindigkeitsgradient $|u'|$ übergeben. Damit ist eine leichte Austauschbarkeit der Materialfunktionen gewährleistet. Es könnte sogar ein nur als Wertetabelle gegebener Zusammenhang (Meßergebnisse) realisiert werden.

— Wahl der Anfangsnäherung:

Aus der Iterationsvorschrift für die Folge der Näherungen $u_n(t)$ geht hervor, daß die Anfangsnäherung (vgl. (2.6))

$$u_0 = \sum_{k=1}^{N(0)} c_{0,k} h_k$$

beliebig gewählt werden kann. Für den Fall der Iteration über das ganze Intervall $[0, T]$ wird

$$u_0 = 0, \quad \text{d. h.} \quad c_{0,k} = 0 \quad \text{für} \quad k = 1, 2, \ldots, N(0)$$

gesetzt. Für den nichtperiodischen Fall existiert aber noch eine andere Möglichkeit, die den Iterationsaufwand wesentlich verringert. Zerlegt man das Gesamtzeitintervall $[0, T]$ in die Teilintervalle $[0, T_1]$, $[T_1, T_2]$, ..., $[T_i, T]$ und wendet die Iterationsvorschrift (2.7) auf jedes Teilintervall an, so erhält man für die folgenden Teilintervalle eine wesentlich bessere Anfangsnäherung, wenn man die Lösung des gerade berechneten Teilintervalls auf das folgende extrapoliert.

— Abbruch der Iteration:

Die Bedingung für den Abbruch der Iteration ergibt sich aus einer Fehlerabschätzung, die für das Fixpunktproblem angegeben werden kann [2]. Danach gilt

$$\int_{R_i}^{R_a} r\big(u_{n+1}(r,t) - u(r,t)\big)^2\, \mathrm{d}r \leqq \frac{a}{2} \left(\frac{q}{1-q}\right)^2 \int_0^T \int_{R_i}^{R_a} r\big(u_n'(r,t) - u_{n+1}'(r,t)\big)^2\, \mathrm{d}r\, \mathrm{d}t\,, \quad n = 0, 1, 2, \ldots. \tag{3.2}$$

Diese Abschätzung gilt strenggenommen nur, wenn die projizierte Lösung $u_N(r, t) = u(r, t)$ ist.

Die Iteration wird beendet, wenn gilt

$$\delta > \frac{q}{1-q} \sqrt{\frac{a}{R_a^2 - R_i^2} \int_0^T \int_{R_i}^{R_a} r \left(\frac{u_n'(r,t) - u_{n+1}'(r,t)}{\max\limits_{r,t}\,(u_{n+1})}\right)^2 \mathrm{d}r\, \mathrm{d}t}\,, \tag{3.3}$$

nach (3.2) ergibt sich daraus,

$$\left(\frac{u_{n+1}(r,t) - u(r,t)}{\max\limits_{r,t}\,(u_{n+1})}\right)^2 < \delta^2$$

gesetzt wurde. Der aktuelle Wert für δ richtet sich nach der geforderten Genauigkeit der Lösung.

— Ergebnisgrößen:

Ist die Abbruchbedingung (3.3) erfüllt, so werden für jeden Zeitpunkt t_i die Geschwindigkeiten $u(r)$ und die dazugehörigen Gradienten $u'(r)$ angegeben. Außerdem wird ein Mengendurchsatz

$$Q = \pi(R_a^2 - R_i^2)\, \bar{u} = 2\pi \int_{R_i}^{R_a} u(r)\, r\, \mathrm{d}r \qquad \bar{u} \text{ — mittlere Geschwindigkeit} \tag{3.4}$$

berechnet. Das bedeutet keinen wesentlichen zusätzlichen Aufwand, da nach (2.6) gilt:

$$Q = 2\pi \sum_{k=1}^{N} c_{n,k} \int_{R_i}^{R_a} h_k(r)\, r\, \mathrm{d}r\,. \tag{3.5}$$

Das Integral in Gl. (3.5) ist aber bereits bekannt (vgl. (2.7)).

4. Ergebnisse und Diskussion

Nichtlinearität der Materialfunktion

Zur Untersuchung des Einflusses wachsender Nichtlinearität der Materialfunktion

$$\tau = \mu(|u'|)\, u'$$

wird das Gesetz von WILLIAMSON (vgl. z. B. [3]) mit

$$\mu = \mu_1 + \frac{\mu_2}{\mu_3 + |u'|} \tag{4.1}$$

herangezogen. In Bild 1 sind Funktionen dieses Typs $f_1 - f_4$ mit wachsender Nichtlinearität dargestellt. Als Grenzfall f_B erhält man das Materialgesetz von BINGHAM, das aber nicht den Voraussetzungen (2.4) genügt. Im folgenden wird das Strömungsverhalten nicht-NEWTONscher Medien unter Verwendung der Materialfunktionen $f_1 - f_4$ untersucht. Die Stoffeigenschaften $\mu_1 = 10^{-5}$ p · s/cm² und $\varrho = 1{,}0$ g/cm³ bleiben dabei konstant. In Bild 2 sind die stationären Geschwindigkeitsprofile für die vier Materialfunktionen bei sonst gleichen Bedingungen aufgetragen. Deutlich ist die Tendenz zur Bildung eines „Kastenprofils" bei wachsender Nichtlinearität der Materialfunktion zu erkennen. Die maximale Geschwindigkeit nimmt ab, während die maximalen Geschwindigkeitsgradienten (hier ca. 100 s⁻¹) etwa gleich bleiben. Die zeitliche Entwicklung eines solchen „Kastenprofils" zeigt Bild 3.

Periodische Druckgradientschwankungen

Für die Untersuchung nicht-NEWTONscher Strömungen bei Anwendung eines periodisch schwankenden Druckgradienten der Form (3.1) spielt das Amplitudenverhältnis

$$A_v = \frac{P_A}{P_0} \quad \text{mit} \quad 0 \leqq A_v \leqq 1$$

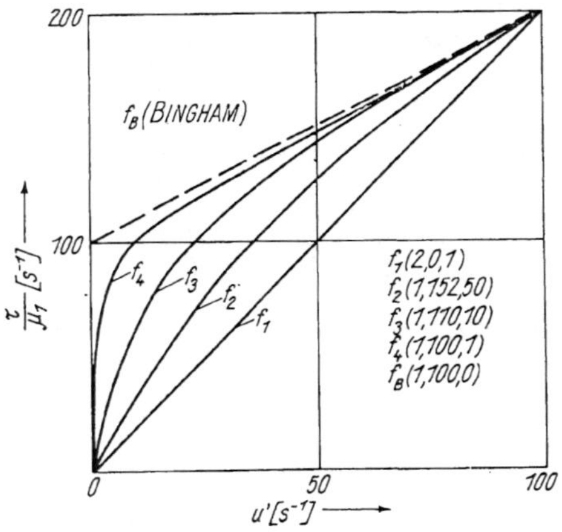

Bild 1. Materialfunktion nach WILLIAMSON $\frac{\tau}{\mu_1} = f u'$ mit $f\left(K, \frac{\mu_2}{\mu_1}, \mu_3\right) = \frac{K\mu}{\mu_1}$ (vgl. Gl. 4.1)

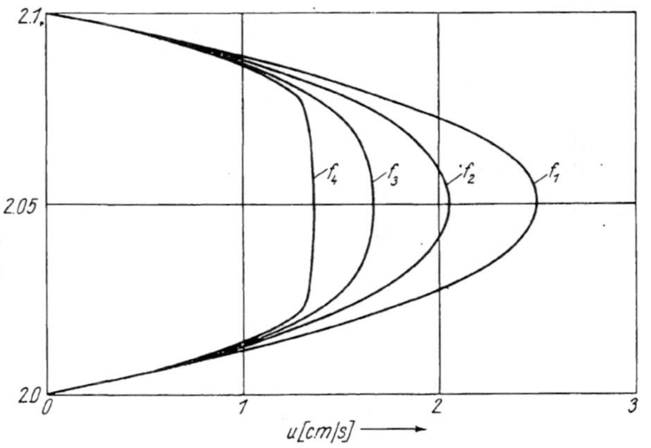

Bild 2. Stationäre Geschwindigkeitsprofile bei Verwendung der Materialfunktionen $f_1 - f_4$ und $P_0 = -0.04$ p/cm³

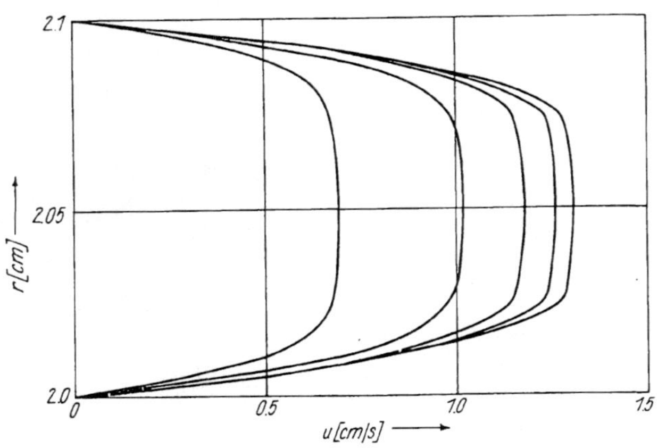

Bild 3. Zeitliche Entwicklung der Geschwindigkeitsprofile für die Materialfunktion f_4 (Zeitabstand 0.06 s, $P_0 = -0.04$ p/cm³)

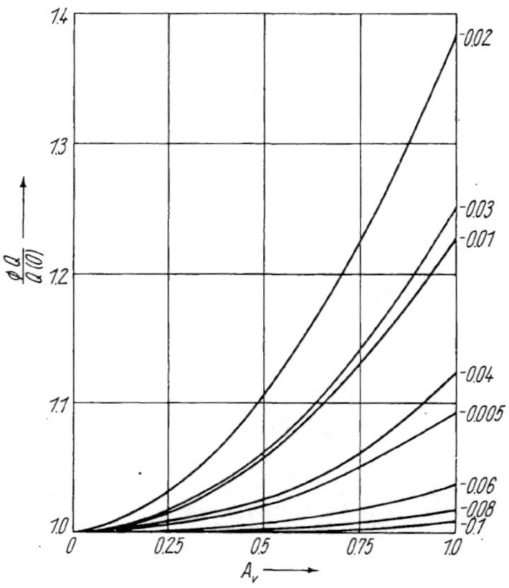

Bild 4. Steigerungsrate des Mengendurchsatzes $\Phi Q(A_v)/Q(0)$ mit Druckgradient P_0 als Parameter (Materialfunktion f_3) ▶

eine wesentliche Rolle. Dabei muß die Lösung der Aufgabe (2.3) für $t \to \infty$ in die Lösung der Aufgabe (2.9) übergehen. Für die Materialfunktion f_4 mit $A_v = 0.5$, $T = 0.3$ s und $P_0 = -0.04$ p/cm^3 waren für $t = 4T$ keine wesentlichen Abweichungen mehr festzustellen.

Für in der Praxis auftretende Probleme ist die Lösung der Aufgabe (2.9) im Hinblick auf eine Beeinflussung des Mengendurchsatzes (vgl. (3.4) von besonderer Bedeutung. In Bild 4 ist die Steigerung des über eine Periode gemittelten Durchsatzes $\varnothing Q$ in Bezug auf den Durchsatz bei konstantem Druckgradient $Q(0)$ über dem Amplitudenverhältnis A_v aufgetragen. Man erkennt einen exponentiellen Anstieg, dessen Intensität stark vom Druckgradienten P_0 abhängt. Diese Abhängigkeit ist in Bild 5 dargestellt. Für ein strömendes nicht-NEWTONsches Medium existiert also ein bestimmter Wert P_0, für den ein überlagerter periodisch schwankender Druckgradient eine maximale Steigerung des Durchsatzes zur Folge hat. Dieser Sachverhalt hängt wesentlich mit der Krümmung der Materialfunktion, d. h. mit der Änderung der effektiven Viskosität des strömenden Materials zusammen. In Bild 6 wird dieser Zusammenhang qualitativ veranschaulicht. Der Verlauf der Krümmung für die Materialfunktion f_3 zeigt ein ausgeprägtes Maximum. Ebenfalls eingezeichnet sind Bereiche der über die Spaltbreite gemittelten Beträge der Geschwindigkeitsgradienten, die für die verschiedenen mittleren Druckgradienten P_0 in Abhängigkeit vom Amplitudenverhältnis auftreten. Das Krümmungsmaximum wird gerade im Fall $P_0 = -0.02$ p/cm^3 überstrichen. Für diesen Druckgradienten ist aber die Möglichkeit der Durchsatzsteigerung besonders groß, wie Bild 5 zeigt. Die Kenntnis des Krümmungsmaximums der Materialfunktion gibt also einen Hinweis für die bestmögliche Durchsatzsteigerung bei Anwendung eines periodisch schwankenden Druckgradienten. Bei weiterer Annäherung der Materialfunktion an den BINGHAM-Fall (vgl. Bild 1) verschiebt sich das Krümmungsmaximum zu kleineren Geschwindigkeitsgradienten hin. Dabei nimmt der Betrag der Krümmung zu.

Das folgende an konkreten praktischen Gegebenheiten orientierte Materialgesetz*)

$$\tau = 344\,|u'|^{-0.7}\,u' \quad [\text{p/cm}^2] \tag{4.2}$$

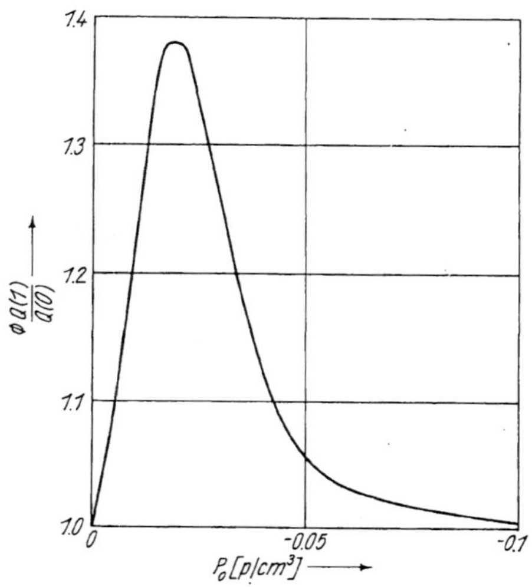

Bild 5. Abhängigkeit der Durchsatzsteigerungsrate vom Druckgradienten P_0 (vgl. Bild 4 für $A_v = 1$)

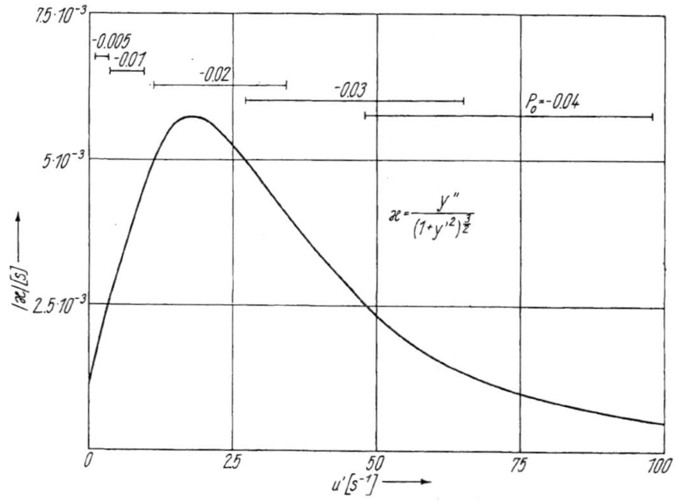

Bild 6. Krümmung der Funktion $y = f_3 u'$ und Gradientenbereiche für verschiedene Druckgradienten P_0

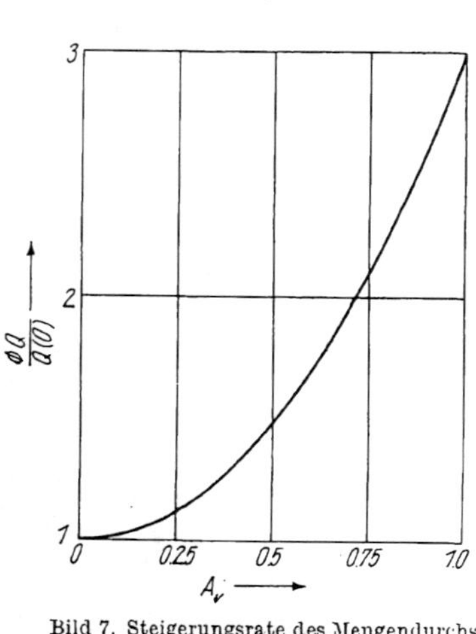

Bild 7. Steigerungsrate des Mengendurchsatzes für das Potenzgesetz (4.2)

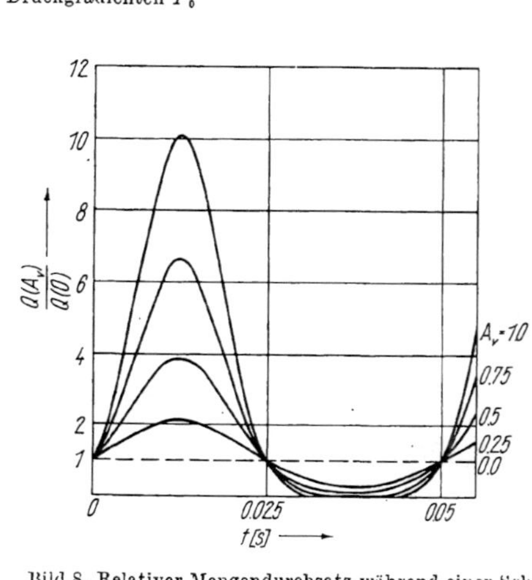

Bild 8. Relativer Mengendurchsatz während einer Schwankungsperiode mit dem Amplitudenverhältnis A_v als Parameter

*) Nach Angaben der Sektion Verfahrenstechnik, Arbeitsgruppe Rheologie der TH Leuna-Merseburg.

wird unter den Bedingungen $\varrho = 0.9$ g/cm³, $P_0' = -5.1$ p/cm³, $R_i = 1.2$ cm, $R_a = 1.32$ cm untersucht. Da das Potenzgesetz (4.2) nicht den Voraussetzungen (2.4) genügt, wird es im Bereich $-0.01\ \mathrm{s}^{-1} < u' < 0.01\ \mathrm{s}^{-1}$ durch eine Gerade mit stetigem Anschluß an die Funktion (4.2) ersetzt. Es wird ferner $|u'| < 10^2$ vorausgesetzt. In Bild 7 ist die Steigerungsrate des Mengendurchsatzes für den o. g. Berechnungsfall dargestellt. Wegen der starken Krümmung von (4.2) ist hier die Steigerungsmöglichkeit besonders groß. Bild 8 zeigt den dazugehörigen zeitlichen Verlauf des Durchsatzes während einer Periode mit dem Amplitudenverhältnis als Parameter. Durch die relativ starke Verringerung der effektiven Zähigkeit bei Vergrößerung der Geschwindigkeiten wird die obere Halbwelle begünstigt. Diese Asymmetrie tritt um so stärker hervor, je größer das Amplitudenverhältnis A_v ist (vgl. Bild 8).

Neben der Abhängigkeit der Durchsatzsteigerung vom Amplitiudenverhältnis und vom mittleren Druckgradient ist auch ein Einfluß der Frequenz der Druckschwankung zu erwarten. Dieser Effekt wird hier nicht untersucht.

5. Schlußbemerkungen

Die Berechnungen wurden an der Rechenanlage BESM 6 des Zentrums für Rechentechnik der AdW der DDR durchgeführt. Die erforderliche Rechenzeit ist stark problemabhängig. Folgende Faktoren sind dafür verantwortlich:

1. Die Kontraktionskonstante $q = \dfrac{M-m}{M+m}$ hängt von der aktuellen Materialfunktion ab (vgl. (2.4)).
2. Die Anzahl $N(n)$ der Koordinatenfunktionen h_k (vgl. 2.6) wird begrenzt. Dieser Wert $N_{\max}$ richtet sich nach der Form der Geschwindigkeitslösung im Ringspalt. Er wird so gebildet, daß die Lösung durch Hinzunahme einer weiteren Koordinatenfunktion nur unwesentlich verändert wird. Für die einzelnen Berechnungsfälle galt $N_{\max} \leqq 15$.
3. Die örtliche und zeitliche Diskretisierung darf die Lösung nicht beeinflussen.
4. Um eine ausreichende Genauigkeit der Lösung zu garantieren, muß gemäß (3.3) ein Wert für δ festgelegt werden, dessen Betrag die Anzahl der erforderlichen Iterationen wesentlich mitbestimmt.

Im Hinblick auf eine breitere Anwendung der angegebenen numerischen Methode soll abschließend bemerkt werden, daß es keine wesentlichen Schwierigkeiten bereitet, den Fall einer nicht-NEWTONschen Strömung im Rohr zu behandeln, dessen durchströmter Querschnitt ein Kreisgebiet ist.

Der Autor dankt den Herren Dr. habil. H. GAJEWSKI und Dr. sc. K. GRÖGER für anregende Diskussionen und wertvolle Hinweise, die zur Entstehung dieser Arbeit beigetragen haben.

6. Literatur

1 GAJEWSKI, H., und GRÖGER, K., Ein Projektions-Iterationsverfahren für Evolutionsgleichungen, Math. Nachr. (erscheint demnächst).

2 GAJEWSKI, H., und GRÖGER, K., Ein Iterationsverfahren für Gleichungen mit einem maximal monotonen und einem stark monotonen LIPSCHITZ-stetigen Operator, Math. Nachr., 68, 307–317 (1975).

3 REHER, E. O., Zur Anwendung der phänomenologischen Reibungsgesetze für nicht-NEMTONsche Flüssigkeiten in der Verfahrenstechnik, Sonderdruck, Wiss. Zeitschr. d. TH Leuna-Merseburg, 9, Heft 3 (1967).

Eingereicht am 17. 9. 1975

Anschrift: Dr. HERMANN WILKE, Berlin, Akademie der Wissenschaften der DDR, Zentralinstitut für Mathematik und Mechanik, 108 Berlin, Mohrenstraße 39, DDR

KLEINE MITTEILUNGEN

ZAMM 56, 499 – 502 (1976)

U. Werner

Turbulente Strömung im mittleren Bereich eines Kanals[1])

Aus der Kontinuitätsgleichung und den Navier-Stokes-Gleichungen ergeben sich für eine inkompressible turbulente Strömung bei Zerlegung der Geschwindigkeit u_i in die mittlere Geschwindigkeit $\bar{u}_i$ und die Schwankungsgeschwindigkeit u'_i und entsprechender Zerlegung des Druckes p in $\bar{p}$ und p'

$$\frac{\partial \bar{u}_i}{\partial x_i} = 0\,, \tag{1}$$

$$\frac{\partial u'_i}{\partial x_i} = 0\,, \tag{2}$$

$$\frac{\partial \bar{u}_i}{\partial t} + \frac{\partial}{\partial x_k}(\bar{u}_k \bar{u}_i + \overline{u'_k u'_i}) = -\frac{1}{\varrho}\frac{\partial \bar{p}}{\partial x_i} + \nu \frac{\partial^2 \bar{u}_i}{\partial x_k \partial x_k}, \tag{3}$$

$$\frac{\partial u'_i}{\partial t} + \frac{\partial}{\partial x_k}(u'_k \bar{u}_i + \bar{u}_k u'_i + u'_k u'_i - \overline{u'_k u'_i}) = -\frac{1}{\varrho}\frac{\partial p'}{\partial x_i} + \nu \frac{\partial^2 u'_i}{\partial x_k \partial x_k}. \tag{4}$$

Die grundlegende Schwierigkeit der Turbulenztheorie zeigt sich darin, daß in diesem System die Anzahl der unbekannten Funktionen größer ist als die Anzahl der verfügbaren Gleichungen. Man ist folglich gezwungen, dem System weitere Gleichungen hinzuzufügen und auf diese Weise das System abzuschließen. Dies gelingt nicht ohne die Verwendung zusätzlicher hypothetischer Annahmen.

Mellor und Herring [1] haben, wobei sie sich auf eine Reihe weiterer Autoren stützen, folgenden Abschluß des DGl-Systems vorgeschlagen. Gleichung (4) wird mit u'_j multipliziert, und von der so entstandenen Gleichung wird der Mittelwert gebildet. Wiederholt man diese Rechnung, wobei die Indizes vertauscht werden, und addiert beide Gleichungen, so ergibt sich bei Berücksichtigung der Kontinuitätsgleichung

$$\left(\frac{\partial}{\partial t} + \bar{u}_k \frac{\partial}{\partial x_k}\right)\overline{u'_i u'_j} = \frac{\partial}{\partial x_k}\left(\nu \frac{\partial}{\partial x_k}\overline{u'_i u'_j} - \overline{u'_k u'_i u'_j} - \frac{1}{\varrho}\overline{p' u'_i}\,\delta_{kj} - \frac{1}{\varrho}\overline{p' u'_j}\,\delta_{ki}\right) + \frac{1}{\varrho}\overline{p'\frac{\partial u'_i}{\partial x_j}} + \frac{1}{\varrho}\overline{p'\frac{\partial u'_j}{\partial x_i}} - \overline{u'_k u'_j}\frac{\partial \bar{u}_i}{\partial x_k} - \overline{u'_k u'_i}\frac{\partial \bar{u}_j}{\partial x_k} - 2\nu\overline{\frac{\partial u'_i}{\partial x_k}\frac{\partial u'_j}{\partial x_k}}. \tag{5}$$

Neben den Schwankungsmomenten 2. Ordnung, deren Bestimmung angestrebt wird, treten in dieser DGl eine Reihe weiterer Momente auf. Letztere sind durch geeignete Abschlußhypothesen zu eliminieren. Mellor und Herring geben dafür folgende Beziehungen zwischen den Momenten 2. Ordnung und den restlichen Momenten an:

$$\frac{1}{\varrho}\overline{p'\frac{\partial u'_i}{\partial x_j}} = -\frac{1}{6}\frac{q}{l_1}\left(\overline{u'_i u'_j} - \frac{1}{3}\delta_{ij} q^2\right) + \frac{1}{2} C q^2\left(\frac{\partial \bar{u}_i}{\partial x_j} + \frac{\partial \bar{u}_j}{\partial x_i}\right),$$

$$\frac{1}{\varrho}\overline{p' u'_i} = -\frac{1}{2} l_2 q \frac{\partial q^2}{\partial x_i},$$

$$\overline{u'_i u'_j u'_k} = -l_3 q\left(\frac{\partial \overline{u'_i u'_j}}{\partial x_k} + \frac{\partial \overline{u'_i u'_k}}{\partial x_j} + \frac{\partial \overline{u'_j u'_k}}{\partial x_i}\right),$$

$$\nu\overline{\frac{\partial u'_i}{\partial x_k}\frac{\partial u'_j}{\partial x_k}} = \frac{1}{3}\frac{q^3}{\Lambda}\delta_{ij}.$$

Dabei ist q^2 die Spur des Korrelationstensors

$$q^2 = \overline{u_1'^2} + \overline{u_2'^2} + \overline{u_3'^2}.$$

l_1, l_2, l_3, Λ sind Proportionalitätsfaktoren mit der Dimension einer Länge.

Mit diesen Abschlußhypothesen liefert (5) die noch fehlenden Gleichungen für die Momente $\overline{u'_i u'_j}$, so daß mit (1), (3), (6) 10 Gleichungen für die 10 Funktionen $\bar{p}$, $\bar{u}_i$, $\overline{u'_i u'_j}$ zur Verfügung stehen:

$$\left(\frac{\partial}{\partial t} + \bar{u}_k \frac{\partial}{\partial x_k}\right)\overline{u'_i u'_j} = \frac{\partial}{\partial x_k}\left[\nu \frac{\partial \overline{u'_i u'_j}}{\partial x_k} + l_3 q\left(\frac{\partial \overline{u'_i u'_j}}{\partial x_k} + \frac{\partial \overline{u'_i u'_k}}{\partial x_j} + \frac{\partial \overline{u'_j u'_k}}{\partial x_i}\right) + \frac{1}{2} l_2 q\left(\delta_{kj}\frac{\partial q^2}{\partial x_i} + \delta_{ki}\frac{\partial q^2}{\partial x_j}\right)\right] - \frac{1}{3}\frac{q}{l_1}\left(\overline{u'_i u'_j} - \frac{1}{3}\delta_{ij} q^2\right) + C q^2\left(\frac{\partial \bar{u}_i}{\partial x_j} + \frac{\partial \bar{u}_j}{\partial x_i}\right) - \overline{u'_k u'_j}\frac{\partial \bar{u}_i}{\partial x_k} - \overline{u'_k u'_i}\frac{\partial \bar{u}_j}{\partial x_k} - \frac{2}{3}\frac{q^3}{\Lambda}\delta_{ij}. \tag{6}$$

Welche Folgerungen sich aus diesem DGl-System für eine konkrete turbulente Strömung ergeben, soll am Fall einer ebenen Kanalströmung betrachtet werden. Die Strömung erfolge zwischen zwei ebenen parallelen Platten, die den Abstand $2h$ voneinander haben. Die Platten sollen so groß sein, daß in guter Näherung angenommen werden kann, daß die mittlere Strömung parallel zu den Begrenzungsebenen verläuft. Weiterhin können sämtliche Mittelwerte als stationär angesehen werden.

Der Beschreibung wird ein kartesisches Koordinatensystem zugrunde gelegt, dessen Ursprung auf der Symmetrieebene des Kanals liegt. Die $x_1 \equiv x$-Achse habe die Richtung der mittleren Geschwindigkeit, die $x_2 \equiv y$-Achse stehe senkrecht auf der Symmetrieebene. Die Platten werden dann durch $y = h$ bzw. $y = -h$ beschrieben. In diesem Koordinatensystem gilt für die mittlere Geschwindigkeit

$$\bar{u}_1 = U(y)\,, \quad \bar{u}_2 = \bar{u}_3 = 0\,.$$

Die Mittelwerte der Momente sind ebenfalls nur von y abhängig; zwei dieser Momente verschwinden aus Symmetriegründen:

$$\overline{u'_i u'_j} = \overline{u'_i u'_j}(y)\,, \quad \overline{u'_1 u'_3} = \overline{u'_2 u'_3} = 0\,.$$

Die Längenmaße l_1, l_2, l_3, Λ werden hier als konstant angesehen. Wenn diese Festlegung auch in unmittelbarer Wandnähe recht zweifelhaft ist, so kann sie doch in Hinblick auf die unten vorgenommene Beschränkung auf den mittleren Kanalbereich als gerechtfertigt angesehen werden. Unter diesen Bedingungen ergibt sich für die Beschreibung der turbulenten Kanalströmung folgendes DGl-System für die Mittelwerte:

$$\frac{\partial \overline{u'_1 u'_2}}{\partial y} = -\frac{1}{\varrho}\frac{\partial \bar{p}}{\partial x} + \nu \frac{\partial^2 U}{\partial y^2}, \tag{3.1}$$

$$\frac{\partial \overline{u'_2 u'_2}}{\partial y} = -\frac{1}{\varrho}\frac{\partial \bar{p}}{\partial y}, \tag{3.2}$$

$$0 = -\frac{1}{\varrho}\frac{\partial \bar{p}}{\partial z}, \tag{3.3}$$

$$0 = (\nu + l_3 q)\frac{\partial^2 \overline{u_1'^2}}{\partial y^2} + l_3 \frac{\partial q}{\partial y}\frac{\partial \overline{u_1'^2}}{\partial y} - \frac{1}{3}\frac{q}{l_1}\left(\overline{u_1'^2} - \frac{1}{3}q^2\right) - 2\overline{u'_1 u'_2}\frac{\partial U}{\partial y} - \frac{2}{3}\frac{q^3}{\Lambda}, \tag{6.11}$$

$$0 = (\nu + 3 l_3 q)\frac{\partial^2 \overline{u_2'^2}}{\partial y^2} + 3 l_3 \frac{\partial q}{\partial y}\frac{\partial \overline{u_2'^2}}{\partial y} + l_2 q \frac{\partial^2 q^2}{\partial y^2} + l_2 \frac{\partial q}{\partial y}\frac{\partial q^2}{\partial y} - \frac{1}{3}\frac{q}{l_1}\left(\overline{u_2'^2} - \frac{1}{3}q^2\right) - \frac{2}{3}\frac{q^3}{\Lambda}, \tag{6.22}$$

[1]) Diese Arbeit entstand im Rahmen eines Forschungsaufenthaltes am ZI für Mathematik und Mechanik der AdW der DDR. Der Verfasser dankt Herrn Dr. G. Seifert für seine wertvollen Hinweise und Anregungen bei der Betreuung.

$$0 = (\nu + l_3 q)\frac{\partial^2 \overline{u_3'^2}}{\partial y^2} + l_3 \frac{\partial q}{\partial y}\frac{\partial \overline{u_3'^2}}{\partial y} - \frac{1}{3}\frac{q}{l_1}\left(\overline{u_3'^2} - \frac{1}{3} q^2\right) - \frac{2}{3}\frac{q^3}{\Lambda}, \tag{6.33}$$

$$0 = (\nu + 2l_3 q)\frac{\partial^2 \overline{u_1' u_2'}}{\partial y^2} + 2l_3 \frac{\partial q}{\partial y}\frac{\partial \overline{u_1' u_2'}}{\partial y} - \frac{1}{3}\frac{q}{l_1}\overline{u_1' u_2'} + Cq^2 \frac{\partial U}{\partial y} - \overline{u_2'^2}\frac{\partial U}{\partial y}. \tag{6.12}$$

Die nicht mehr angeführten Gleichungen sind identisch erfüllt. Der mittlere Druck $\bar{p}$ kann aus diesem System noch eliminiert werden, denn die Gleichungen (3) liefern sofort

$$\frac{1}{\varrho}\bar{p} = -\overline{u_2' u_2'}(y) + \dot{a}x + b\,.$$

Da die Anwendbarkeit der von MELLOR und HERRING benutzten Abschlußhypothesen auf Gebiete in Wandnähe nicht gesichert erscheint, soll das vorgelegte DGl-System hier nur für den zentralen Bereich des Kanals gelöst werden. Für die Umgebung der Kanalmitte sind zwei vereinfachende Annahmen gerechtfertigt, die eine näherungsweise Lösung des DGl-Systems gestatten:

1. Versuchsergebnisse zeigen, daß das mittlere Geschwindigkeitsprofil bei turbulenter Strömung wesentlich von dem einer laminaren Strömung abweicht. Die mittlere Geschwindigkeit $U(y)$ steigt in Wandnähe stark an, um dann im mittleren Bereich des Kanals fast konstant zu bleiben. (COMTE-BELLOT [2]: Bei $y = h/2$ weicht $U(y)$ nur um 7% von der Maximalgeschwindigkeit U_0 in der Kanalmitte ab, bei $y = h/4$ weniger als 3%.) Dies rechtfertigt die Annahme

   $U(y) = \text{const.} = U_0\,.$

2. Die Zähigkeit macht sich besonders in Wandnähe bemerkbar. In der Kanalmitte ist sie von untergeordneter Bedeutung, weshalb ν gegen $l_3 q$ vernachlässigt werden kann. Es ist zu erwarten, daß dies besonders bei großen REYNOLDSschen Zahlen gerechtfertigt ist.

Im folgenden wird zu dimensionslosen Größen übergegangen, wobei alle Längen auf die halbe Kanalbreite h, die Geschwindigkeiten auf eine charakteristische Geschwindigkeit u_0 (möglich ist etwa die Maximalgeschwindigkeit U_0) bezogen werden:

$$\frac{y}{h} = \eta\,, \quad \frac{l_i}{h} = L_i\,, \quad \frac{\Lambda}{h} = \lambda\,,$$

$$\frac{\overline{u_1'^2}}{u_0^2} = u\,, \quad \frac{\overline{u_2'^2}}{u_0^2} = v\,, \quad \frac{\overline{u_3'^2}}{u_0^2} = w\,, \quad \frac{q}{u_0} = Q\,.$$

Unter den obigen beiden Annahmen vereinfacht sich das DGl-System zu (sämtliche Ableitungen sind nach η zu bilden)

$$\left(\frac{\overline{u_1' u_2'}}{u_0^2}\right)' = -\frac{h}{u_0^2}\frac{1}{\varrho}\frac{\partial \bar{p}}{\partial x}, \tag{3.1*}$$

$$L_3 Q u'' + L_3 Q' u' - \frac{1}{3L_1} Q\left(u - \frac{1}{3} Q^2\right) - \frac{2}{3}\frac{1}{\lambda} Q^3 = 0\,, \tag{6.11*}$$

$$3L_3 Q v'' + 3L_3 Q' v' + L_2 Q (Q^2)'' + L_2 Q' (Q^2)' - \frac{1}{3L_1} Q\left(v - \frac{1}{3} Q^2\right) - \frac{2}{3}\frac{1}{\lambda} Q^3 = 0\,, \tag{6.22*}$$

$$L_3 Q w'' + L_3 Q' w' - \frac{1}{3L_1} Q\left(w - \frac{1}{3} Q^2\right) - \frac{2}{3}\frac{1}{\lambda} Q^3 = 0, \tag{6.33*}$$

$$2L_3 Q\left(\frac{\overline{u_1' u_2'}}{u_0^2}\right)'' + 2L_3 Q'\left(\frac{\overline{u_1' u_2'}}{u_0^2}\right)' - \frac{1}{3L_1} Q \frac{\overline{u_1' u_2'}}{u_0^2} = 0\,. \tag{6.12*}$$

Weil mit $U(y) = U_0$ eine der Lösungsfunktionen bereits festgelegt wurde und $\frac{1}{\varrho}\frac{\partial \bar{p}}{\partial x} = a$ als eine vorgegebene Konstante zu betrachten ist, ist dieses System aus 5 Gleichungen für die Funktionen $\frac{\overline{u_1' u_2'}}{u_0^2}$, u, v, w, Q überbestimmt, denn wegen

$$Q^2 = u + v + w \tag{7}$$

sind diese Funktionen nicht unabhängig voneinander. Um diese Schwierigkeit zu umgehen, wird (7) bei der Lösung des Systems nicht berücksichtigt, sondern $Q(\eta)$ als selbständige Funktion behandelt. Anschließend ist nachzuprüfen, inwieweit die Abhängigkeit (7) erfüllt ist.

Durch die für den zentralen Kanalbereich angenommenen Vereinfachungen sind die Gleichungen (6*) derart beschaffen, daß nicht mehr sämtliche Längenmaße bestimmt werden können. Durch Multiplikation etwa mit L_1 treten sie nur noch in den Kombinationen

$$\alpha = 6L_1 L_3, \quad \frac{L_1}{\lambda}, \quad \frac{L_2}{L_3} \tag{8}$$

auf.

Naturgemäß stehen jetzt die Haftbedingungen an den Kanalwänden als Randbedingungen nicht zur Verfügung. Sie werden durch Anfangsbedingungen in der Kanalmitte $\eta = 0$ ersetzt. Aus Symmetriegründen müssen $\overline{u_1'^2}$, $\overline{u_2'^2}$, $\overline{u_3'^2}$ gerade Funktionen, $\overline{u_1' u_2'}$ dagegen ungerade sein:

$$u'(0) = 0\,, \quad v'(0) = 0\,, \quad w'(0) = 0 \quad \frac{\overline{u_1' u_2'}(0)}{u_0^2} = 0\,.$$

Die fehlenden Anfangswerte $u(0)$, $v(0)$, $w(0)$ können nur aus Meßwerten gewonnen werden.

(3.1*) besitzt bei Berücksichtigung der Anfangsbedingung die Lösung

$$\frac{\overline{u_1' u_2'}}{u_0^2} = -\frac{h}{u_0^2}\frac{1}{\varrho}\frac{\partial \bar{p}}{\partial x}\eta\,. \tag{9}$$

Hiermit ergibt (6.12*)

$$2L_3 Q' - \frac{1}{3L_1}\eta Q = 0$$

mit der Lösung

$$Q(\eta) = Q(0)\,e^{\frac{\eta^2}{2\alpha}}, \tag{10}$$

wobei α gemäß (8) festgelegt ist.

Nach Einsetzen dieser Funktion in (6.11*)—(6.33*) ergibt sich jeweils eine DGl für u, v, w. Diese sind nicht mehr gekoppelt. Außerdem sind sie linear, so daß ihre allgemeinen Lösungen aus partikulären Lösungen superponiert werden können. (6.11*) geht in die inhomogene lineare DGL

$$L_3 u'' + \frac{L_3}{\alpha}\eta u' - \frac{1}{3L_1} u = \left(\frac{2}{3\lambda} - \frac{1}{9L_1}\right) Q^2(0)\,e^{\frac{\eta^2}{\alpha}}$$

über, welche

$$u_1(\eta) = \eta^2 + \alpha$$

als Partikulärlösung besitzt. Daraus ergibt sich dann

$$u(\eta) = (\eta^2 + \alpha)\left[C_1 + C_2 J(\eta) + \frac{2}{3}\left(\frac{1}{\lambda L_3} - \frac{1}{\alpha}\right) \times Q^2(0)\left\{J(\eta)\int_0^\eta (\eta^2 + \alpha)\,e^{\frac{3\eta^2}{2\alpha}}\,d\eta - \int_0^\eta J(\eta)\,(\eta^2 + \alpha)\,e^{\frac{3\eta^2}{2\alpha}}\,d\eta\right\}\right] \tag{11}$$

mit

$$J(\eta) = \int_0^\eta \frac{e^{-\frac{\eta^2}{2\alpha}}}{(\eta^2 + \alpha)^2}\,d\eta$$

als allgemeine Lösung. Die Anfangsbedingungen verlangen die Konstantenfestlegung

$$C_1 = \frac{u(0)}{\alpha}, \quad C_2 = 0\,.$$

Die DGl (6.33*) stimmt vollständig mit (6.11*) überein. Ihre Lösung unterscheidet sich folglich von (11) nur im Anfangswert $w(0)$:

$$w(\eta) = (\eta^2 + \alpha)\left[\frac{w(0)}{\alpha} + \frac{2}{3}\left(\frac{1}{\lambda L_3} - \frac{1}{\alpha}\right) \times \right.$$
$$\left. \times\, Q^2(0)\left\{J(\eta)\int_0^\eta (\eta^2 + \alpha)\, e^{\frac{3\eta^2}{2\alpha}}\, d\eta - \int_0^\eta J(\eta)(\eta^2 + \alpha) e^{\frac{3\eta^2}{2\alpha}}\, d\eta\right\}\right]. \tag{12}$$

Die Gleichung (6.22*) für die letzte der gesuchten Funktionen nimmt nach Einsetzen von (10) die Gestalt

$$3L_3 v'' + \frac{3L_3}{\alpha}\eta v' - \frac{1}{3L_1} v = \left(\frac{2}{3\lambda} - \frac{1}{9L_1} - \frac{2L_2}{\alpha} - \frac{6L_2}{\alpha^2}\eta^2\right) Q^2(0)\, e^{\frac{\eta^2}{\alpha}}$$

an. Die zugehörige homogene DGl läßt sich nicht durch Quadraturen lösen. Ein Reihenansatz, bei dem die Anfangsbedingung $v'(0) = 0$ von vornherein berücksichtigt wird, liefert als Lösungsfunktion

$$v_1(\eta) = 1 + \sum_{n=1}^{\infty} \frac{(-1)^n}{(3\alpha)^n} \frac{(-2)\cdot 4\cdot 10 \cdots (6n-8)}{(2n)!}\, \eta^{2n} = R(\eta)\,.$$

Mit Hilfe dieser Reihe läßt sich die allgemeine Lösung darstellen:

$$v(\eta) = R(\eta)\left[C_1 + C_2 K(\eta) + \frac{1}{3L_3} \times \right.$$
$$\times\, Q^2(0)\left\{K(\eta)\int_0^\eta \left(\frac{2}{3\lambda} - \frac{1}{9L_1} - \frac{2L_2}{\alpha} - \frac{6L_2}{\alpha^2}\eta^2\right) R(\eta)\, e^{\frac{3\eta^2}{2\alpha}}\, d\eta - \right.$$
$$\left.\left. - \int_0^\eta K(\eta)\left(\frac{2}{3\lambda} - \frac{1}{9L_1} - \frac{2L_2}{\alpha} - \frac{6L_2}{\alpha^2}\eta^2\right) R(\eta)\, e^{\frac{3\eta^2}{2\alpha}}\, d\eta\right\}\right] \tag{13}$$

mit

$$K(\eta) = \int_0^\eta \frac{e^{-\frac{\eta^2}{2\alpha}}}{R^2(\eta)}\, d\eta\,.$$

Die Anfangsbedingungen liefern

$$C_1 = v(0)\,, \quad C_2 = 0\,.$$

Die erhaltenen Lösungsfunktionen (10), (11), (12), (13) erfüllen offensichtlich die Bedingung (7) nicht. Inwieweit zumindest eine teilweise Übereinstimmung vorliegt, kann über die Reihenentwicklungen der einzelnen Funktionen ermittelt werden. Die noch verfügbaren Längenmaße sollen dabei so festgelegt werden, daß (7) bis zu Gliedern möglichst hoher Ordnung erfüllt ist. Entwickelt man u, v, w, Q^2 in Potenzreihen und setzt diese in (7) ein, so ergibt sich durch Koeffizientenvergleich: Die Absolutglieder erfüllen die geforderte Beziehung

$$Q^2(0) = u(0) + v(0) + w(0)\,. \tag{14}$$

Die Koeffizienten von η^2 (es treten nur gerade Potenzen von η auf) liefern

$$\frac{1}{\alpha} Q^2(0) = \frac{u(0)}{\alpha} + \frac{w(0)}{\alpha} + \frac{2}{3}\left(\frac{1}{\lambda L_3} - \frac{1}{\alpha}\right) Q^2(0) +$$
$$+ \frac{1}{3}\frac{v(0)}{\alpha} + \frac{1}{3}\left(2\frac{L_1}{\lambda} - \frac{L_2}{L_3} - \frac{1}{3}\right)\frac{1}{\alpha} Q^2(0)\,,$$

woraus sich unter Beachtung von (14)

$$14\frac{L_1}{\lambda} - \frac{L_2}{L_3} - 2\frac{v(0)}{Q^2(0)} - \frac{7}{3} = 0 \tag{15}$$

ergibt. Eine Übereinstimmung der Koeffizienten bei η^2 ist also nur dann zu erreichen, wenn zwischen den Längenmaßen eine Abhängigkeit gemäß (15) festgelegt wird. Analog liefern die Koeffizienten bei η^4

$$38\frac{L_1}{\lambda} - 10\frac{L_2}{L_3} - 2\frac{v(0)}{Q^2(0)} - \frac{100}{3} = 0\,. \tag{16}$$

Das Gleichungssystem (15), (16) besitzt die Lösung

$$\frac{L_1}{\lambda} = \frac{1}{51}\left(9\frac{v(0)}{Q^2(0)} - 5\right), \quad \frac{L_2}{L_3} = \frac{1}{51}\left(24\frac{v(0)}{Q^2(0)} - 189\right). \tag{17}$$

Eine Übereinstimmung der Koeffizienten bei η^6 kann nicht mehr erreicht werden. Für $\alpha = 6L_1L_3$ ergibt sich aus dem Vergleich keine Bedingung.

Bricht man also die Reihenentwicklung für u, v, w, Q^2 nach den Gliedern mit η^4 ab, so ergeben sich Näherungslösungen des vorgelegten DGl-Systems, die der Bedingung (7) genügen:

$$\frac{\overline{u_1'^2}(\eta)}{u_0^2} = \frac{1}{u_0^2}\left\{\overline{u_1'^2}(0) + \left[\overline{u_1'^2}(0) + \frac{3}{17}\left(2\overline{u_2'^2}(0) - 3q^2(0)\right)\right]\frac{\eta^2}{\alpha} + \right.$$
$$\left. + \frac{1}{34}\left[2\overline{u_2'^2}(0) - 3q^2(0)\right]\frac{\eta^4}{\alpha^2}\right\},$$

$$\frac{\overline{u_2'^2}(\eta)}{u_0^2} = \frac{1}{u_0^2}\left\{\overline{u_2'^2}(0) + \frac{1}{17}\left[5\overline{u_2'^2}(0) + 18q^2(0)\right]\frac{\eta^2}{\alpha} + \right.$$
$$\left. + \frac{1}{34}\left[-4\overline{u_2'^2}(0) + 23q^2(0)\right]\frac{\eta^4}{\alpha^2}\right\},$$

$$\frac{\overline{u_3'^2}(\eta)}{u_0^2} = \frac{1}{u_0^2}\left\{\overline{u_3'^2}(0) + \left[\overline{u_3'^2}(0) + \frac{3}{17}\left(2\overline{u_2'^2}(0) - 3q^2(0)\right)\right]\frac{\eta^2}{\alpha} + \right.$$
$$\left. + \frac{1}{34}\left[2\overline{u_2'^2}(0) - 3q^2(0)\right]\frac{\eta^4}{\alpha^2}\right\},$$

$$\frac{q^2(\eta)}{u_0^2} = \frac{q^2(0)}{u_0^2}\left\{1 + \frac{1}{\alpha}\eta^2 + \frac{1}{2\alpha^2}\eta^4\right\}.$$

Zusammen mit (9) liegen damit sämtliche gesuchten Schwankungsmomente 2. Ordnung vor.

Die noch offenen Konstanten sind aus Meßwerten zu ermitteln. Die von Comte-Bellot [2] durchgeführten Messungen einer turbulenten Strömung in einem ebenen Kanal (Re = 120000; $h = 9$ cm; als Bezugsgeschwindigkeit u_0 wurde die Wandschubspannungsgeschwindigkeit $u_f = 80$ cm/sec verwendet) ergeben im Bereich der ausgebildeten Turbulenz

$$\frac{1}{\varrho}\frac{\partial \bar{p}}{\partial x} = -755\,\frac{\text{cm}}{\text{sec}^2}\,,$$

$$\frac{\overline{u_1'^2}(0)}{u_f^2} = 0{,}9\,, \quad \frac{\overline{u_2'^2}(0)}{u_f^2} = 0{,}5\,, \quad \frac{\overline{u_3'^2}(0)}{u_f^2} = 0{,}45\,.$$

Für die turbulente Reibung (9) ergibt sich hieraus

$$\frac{\overline{u_1'u_2'}}{u_f^2} = 1{,}06\eta\,.$$

Zur Bestimmung von α wird nach der Methode der kleinsten Quadrate eine Annäherung von $q^2(\eta)/u_f^2$ an die Meßwerte durchgeführt, die $\alpha = 0{,}25$ liefert. Für die Längenmaße ergeben sich aus diesen Werten

$$L_1L_3 = 0{,}04\,, \quad \frac{L_1}{\lambda} = -0{,}05\,, \quad \frac{L_2}{L_3} = -3{,}4\,.$$

Zwei der Längenmaße sind positiv, zwei negativ. Da bei den von Mellor und Herring vorgeschlagenen Abschlußhypothesen die einzelnen Momente bereits mit Vorzeichen versehen wurden und daher naturgemäß nur positive Längenmaße zu erwarten waren, ist dieses Ergebnis etwas verwunderlich. Allerdings herrscht in der Literatur keine Einmütigkeit in bezug auf die Vorzeichen der Korrelationsfunktionen. Ob hierin oder in den vorgenommenen Vereinfachungen für die Strömung in der Kanalmitte der Grund für die unterschiedlichen Vorzeichen zu suchen ist, muß hier unbeantwortet bleiben. Die erhaltenen Funktionen zeigen eine gute Übereinstimmung mit den gemessenen Werten, was für die Rechtfertigung der Näherungsvereinbarungen spricht.

Literatur

1 Mellor, G. L., Herring, H. J., A Survey of the Mean Turbulent Field Closure Models, AIAA Journal 11, Nr. 5 (1973).
2 Ж. Конт-Белло (G. Comte-Bellot), Турбулентное течение в канале с параллельными стенками, Изд. Мир, Москва 1968 (Übersetzung aus dem Französischen).

Eingereicht am 23. 7. 1975

Anschrift: Dr. Udo Werner, Ingenieurhochschule Köthen, Abt. Mathematik, DDR-437 Köthen, Bernburger Straße 52—57, DDR

ZAMM 56, 502 – 503 (1976)

Jean Paul Boehler

Sur la loi de comportement d'un milieu orthotrope de révolution en état de déformation plane

A partir des théorèmes de représentations de Wang, ([1] à [3]), nous avons donné précédemment [4] la forme la plus générale de la loi de comportement d'un milieu orthotrope de révolution, lorsque cette loi s'exprime par une relation explicite entre deux tenseurs symétriques du second ordre ($\boldsymbol{T}$ et $\boldsymbol{E}$). L'état de déformation plane présentant un intérêt particulier pour les applications, nous étudions ici les restrictions qu'entraîne cette condition. La loi de comportement s'exprime dans ce cas par une forme quasilinéaire (forme linéaire par rapport aux tenseurs générateurs de la représentation de la loi).

1. Relations tensorielles

Soit un milieu orthotrope de révolution repéré dans les axes orthonormés $(\mathfrak{v}_1, \mathfrak{v}_2, \mathfrak{v}_3)$, $\mathfrak{v}_3$ étant l'axe de révolution de l'orthotropie. La loi de comportement de ce milieu admet comme représentation ([1]):

$$\boldsymbol{T} = \alpha_0 \boldsymbol{I} + \alpha_1 \boldsymbol{M} + \alpha_2 \boldsymbol{E} + \\ + \alpha_3(\boldsymbol{ME} + \boldsymbol{EM}) + \alpha_4 \boldsymbol{E}^2 + \alpha_5(\boldsymbol{ME}^2 + \boldsymbol{E}^2\boldsymbol{M}) , \tag{1}$$
$$\alpha_i = \alpha_i(\mathrm{tr}\, \boldsymbol{E},\ \mathrm{tr}\, \boldsymbol{E}^2,\ \mathrm{tr}\, \boldsymbol{E}^3,\ \mathrm{tr}\, \boldsymbol{ME},\ \mathrm{tr}\, \boldsymbol{ME}^2) , \quad \boldsymbol{M} = \mathfrak{v}_3 \otimes \mathfrak{v}_3 .$$

Dans le cas du comportement élastique par exemple, $\boldsymbol{T}$ représente le tenseur des contraintes σ_{ij} et $\boldsymbol{E}$ le tenseur des déformations ε_{ij}; dans le cas du comportement visqueux, $\boldsymbol{E}$ représente le tenseur des vitesses de déformations.

Supposons que le milieu soit en état de déformation plane dans le plan $(\mathfrak{v}_2, \mathfrak{v}_3)$ contenant l'axe privilégié $\mathfrak{v}_3$. Les conditions:

$$\varepsilon_{11} = \varepsilon_{31} = \varepsilon_{12} = 0 \tag{2}$$

de cette déformation plane donnent à partir de (1):

$$\sigma_{11} = \alpha_0 , \qquad \sigma_{31} = \sigma_{12} = 0 . \tag{3}$$

Soient les tenseurs à deux dimensions:

$$\mathfrak{I} = \begin{vmatrix} 1 & 0 \\ 0 & 1 \end{vmatrix}, \quad \mathfrak{M} = \begin{vmatrix} 0 & 0 \\ 0 & 1 \end{vmatrix}, \quad \mathfrak{E} = \begin{vmatrix} \varepsilon_{22} & \varepsilon_{23} \\ \varepsilon_{23} & \varepsilon_{33} \end{vmatrix}, \quad \mathfrak{T} = \begin{vmatrix} \sigma_{22} & \sigma_{23} \\ \sigma_{23} & \sigma_{33} \end{vmatrix} \tag{4}$$

ainsi définis dans les axes $(\mathfrak{v}_2, \mathfrak{v}_3)$.

Les conditions (2) et les particularités des tenseurs $\boldsymbol{M}$ et $\mathfrak{M}$ entraînent:

$$\mathrm{tr}\, \boldsymbol{E}^p = \mathrm{tr}\, \mathfrak{E}^p , \quad \mathrm{tr}\, \boldsymbol{ME}^p = \mathrm{tr}\, \mathfrak{M}\mathfrak{E}^p , \tag{5}$$

où p est un entier positif.

La loi de comportement (1) peut alors se décomposer en (3) et:

$$\mathfrak{T} = \alpha_0 \mathfrak{I} + \alpha_1 \mathfrak{M} + \alpha_2 \mathfrak{E} + \\ + \alpha_3(\mathfrak{M}\mathfrak{E} + \mathfrak{E}\mathfrak{M}) + \alpha_4 \mathfrak{E}^2 + \alpha_5(\mathfrak{M}\mathfrak{E}^2 + \mathfrak{E}^2\mathfrak{M}) , \tag{6}$$
$$\alpha_i = \alpha_i\,(\mathrm{tr}\, \mathfrak{E},\ \mathrm{tr}\, \mathfrak{E}^2,\ \mathrm{tr}\, \mathfrak{E}^3,\ \mathrm{tr}\, \mathfrak{M}\mathfrak{E},\ \mathrm{tr}\, \mathfrak{M}\mathfrak{E}^2) .$$

La forme (6) se réduit par l'application du théorème de Cayley-Hamilton:

$$\mathfrak{m}^2 - (\mathrm{tr}\, \mathfrak{m})\, \mathfrak{m} + \tfrac{1}{2}\, [(\mathrm{tr}\, \mathfrak{m})^2 - \mathrm{tr}\, \mathfrak{m}^2]\, \mathfrak{I} = 0 \tag{7}$$

et la généralisation de ce théorème ([5]):

$$\mathfrak{m}\mathfrak{n} + \mathfrak{n}\mathfrak{m} - (\mathrm{tr}\, \mathfrak{n})\mathfrak{m} - (\mathrm{tr}\, \mathfrak{m})\mathfrak{n} + (\mathrm{tr}\, \mathfrak{m}\ \mathrm{tr}\, \mathfrak{n} - \mathrm{tr}\, \mathfrak{m}\mathfrak{n})\, \mathfrak{I} = 0 \tag{8}$$

pour des tenseurs $\mathfrak{m}$ et $\mathfrak{n}$ symétriques du second ordre à deux dimensions.

2. Réduction du nombre des invariants indépendants

Le théorème (7) donne:

$$\mathrm{tr}\, \mathfrak{E}^3 = \tfrac{3}{2}\, \mathrm{tr}\, \mathfrak{E}\ \mathrm{tr}\, \mathfrak{E}^2 - \tfrac{1}{2}\, (\mathrm{tr}\, \mathfrak{E})^3 . \tag{9}$$

En posant $\mathfrak{m} = \mathfrak{M}$ et $\mathfrak{n} = \mathfrak{E}$ dans (8), en multipliant à gauche par $\mathfrak{E}$ et en prenant la trace, il vient:

$$\mathrm{tr}\, \mathfrak{M}\mathfrak{E}^2 = \mathrm{tr}\, \mathfrak{E}\ \mathrm{tr}\, \mathfrak{M}\mathfrak{E} + \tfrac{1}{2}\, [\mathrm{tr}\, \mathfrak{E}^2 - (\mathrm{tr}\, \mathfrak{E})^2] . \tag{10}$$

Dans la représentation (6), il reste comme invariants indépendants:

$$\mathrm{tr}\, \mathfrak{E},\ \mathrm{tr}\, \mathfrak{E}^2,\ \mathrm{tr}\, \mathfrak{M}\mathfrak{E} . \tag{11}$$

Comme nous pouvons identifier les tenseurs $\mathfrak{M}$ et $\boldsymbol{M}$ et les tenseurs $\mathfrak{E}$ et $\boldsymbol{E}$ dans le cas de la déformation plane, les arguments de α_0 dans (3) se réduisent également à (11).

3. Réduction du nombre des générateurs

Le théorème (7) donne:

$$\mathfrak{E}^2 = \tfrac{1}{2}\, [\mathrm{tr}\, \mathfrak{E}^2 - (\mathrm{tr}\, \mathfrak{E})^2]\, \mathfrak{I} + (\mathrm{tr}\, \mathfrak{E})\, \mathfrak{E} . \tag{12}$$

En posant $\mathfrak{m} = \mathfrak{M}$ et $\mathfrak{n} = \mathfrak{E}$ dans (8), il vient:

$$\mathfrak{M}\mathfrak{E} + \mathfrak{E}\mathfrak{M} = (\mathrm{tr}\, \mathfrak{M}\mathfrak{E} - \mathrm{tr}\, \mathfrak{E})\, \mathfrak{I} + (\mathrm{tr}\, \mathfrak{E})\, \mathfrak{M} + \mathfrak{E} . \tag{13}$$

En posant $\mathfrak{m} = \mathfrak{M}$ et $\mathfrak{n} = \mathfrak{E}^2$ dans (8), il vient en tenant compte de (10):

$$\mathfrak{M}\mathfrak{E}^2 + \mathfrak{E}^2\mathfrak{M} \\ = [\mathrm{tr}\, \mathfrak{E}\ \mathrm{tr}\, \mathfrak{M}\mathfrak{E} - (\mathrm{tr}\, \mathfrak{E})^2]\, \mathfrak{I} + (\mathrm{tr}\, \mathfrak{E})\, \mathfrak{E} + (\mathrm{tr}\, \mathfrak{E}^2)\, \mathfrak{M} . \tag{14}$$

Les générateurs $\mathfrak{E}^2$, $\mathfrak{M}\mathfrak{E} + \mathfrak{E}\mathfrak{M}$ et $\mathfrak{M}\mathfrak{E}^2 + \mathfrak{E}^2\mathfrak{M}$ de (6) s'expriment donc linéairement en fonction des générateurs indépendants:

$$\mathfrak{I},\ \mathfrak{E},\ \mathfrak{M} \tag{15}$$

à l'aide des invariants (11).

4. Loi de comportement

En utilisant (9), (10), (12), (13) et (14) dans (6), nous obtenons:

$$\mathfrak{T} = \{\alpha_0 + (\mathrm{tr}\, \mathfrak{M}\mathfrak{E} - \mathrm{tr}\, \mathfrak{E})\, \alpha_3 + \tfrac{1}{2}\, [\mathrm{tr}\, \mathfrak{E}^2 - (\mathrm{tr}\, \mathfrak{E})^2]\, \alpha_4 + \\ [\mathrm{tr}\, \mathfrak{E}\ \mathrm{tr}\, \mathfrak{M}\mathfrak{E} - (\mathrm{tr}\, \mathfrak{E})^2]\, \alpha_5\}\, \mathfrak{I} + \\ \{\alpha_1 + (\mathrm{tr}\, \mathfrak{E})\, \alpha_3 + (\mathrm{tr}\, \mathfrak{E}^2)\, \alpha_5\}\, \mathfrak{M} + \\ \{\alpha_2 + \alpha_3 + (\alpha_4 + \alpha_5)\, \mathrm{tr}\, \mathfrak{E}\}\, \mathfrak{E} , \tag{16}$$
$$\alpha_i = \alpha_i\, (\mathrm{tr}\, \mathfrak{E}, \mathrm{tr}\, \mathfrak{E}^2,\ \mathrm{tr}\, \mathfrak{M}\mathfrak{E}) .$$

La loi (1) admet donc comme représentation dans le cas d'une déformation plane dans le plan $(\mathfrak{v}_2, \mathfrak{v}_3)$:

$$\sigma_{11} = \varphi_0 , \qquad \sigma_{31} = \sigma_{12} = 0 , \\ \mathfrak{T} = \varphi_1 \mathfrak{I} + \varphi_2 \mathfrak{M} + \varphi_3 \mathfrak{E} , \\ \varphi_i = \varphi_i\, (\mathrm{tr}\, \mathfrak{E},\ \mathrm{tr}\, \mathfrak{E}^2,\ \mathrm{tr}\, \mathfrak{M}\mathfrak{E}) , \tag{17}$$

où les fonctions scalaires φ_i sont définies à partir des fonctions α_i par les relations (3) et (16).

Nous obtenons une forme linéaire par rapport aux tenseurs générateurs $\mathfrak{I}$, $\mathfrak{M}$ et $\mathfrak{E}$.

Dans le cas d'un milieu orthotrope tel que le plan de déformation $(\mathfrak{v}_2, \mathfrak{v}_3)$ est un plan de symétrie du milieu, l'étude est similaire et on aboutit à la même forme (17) mais avec d'autres fonctions φ_i.

5. Exemple: élasticité infinitésimale linéaire

Le comportement élastique infinitésimale linéaire s'exprime classiquement par:

$$\boldsymbol{T} = \boldsymbol{A}\boldsymbol{E}\,, \tag{18}$$

où $\boldsymbol{A}$ est un tenseur du quatrième ordre. Les composantes de $\boldsymbol{A}$ dans les axes $(\mathfrak{v}_1, \mathfrak{v}_2, \mathfrak{v}_3)$ s'expriment en fonction des cinq constantes d'anisotropie du milieu: $\mathfrak{a}$, $\mathfrak{c}$, $\mathfrak{d}$, $\mathfrak{f}$ et $\mathfrak{g}$. Nous avons établi dans [4] le passage de la forme linéaire de la relation (1) à la relation (18).

Dans le cas de la déformation plane dans le plan $(\mathfrak{v}_2, \mathfrak{v}_3)$, la relation (18) donne:

$$\begin{aligned} \sigma_{11} &= \mathfrak{f}\varepsilon_{22} + \mathfrak{d}\varepsilon_{33}\,, & \sigma_{23} &= 2\mathfrak{g}\varepsilon_{23}\,, \\ \sigma_{22} &= \mathfrak{a}\varepsilon_{22} + \mathfrak{d}\varepsilon_{33}\,, & \sigma_{31} &= 0\,, \\ \sigma_{33} &= \mathfrak{d}\varepsilon_{22} + \mathfrak{c}\varepsilon_{33}\,, & \sigma_{12} &= 0\,. \end{aligned} \tag{19}$$

Les relations (19) doivent s'obtenir à partir de la forme linéaire en $\mathfrak{E}$ de la loi (17): φ_0, φ_1 et φ_2 fonctions linéaires en tr $\mathfrak{E}$ et tr $\mathfrak{M}\mathfrak{E}$; φ_3 constante.

En utilisant les résultats de [4], nous obtenons bien:

$$\begin{aligned} \varphi_0 &= \mathfrak{h}\,\mathrm{tr}\,\mathfrak{E} + (\mathfrak{d} - \mathfrak{f})\,\mathrm{tr}\,\mathfrak{M}\mathfrak{E}\,, \\ \varphi_1 &= (\mathfrak{a} - 2\mathfrak{g})\,\mathrm{tr}\,\mathfrak{E} + (\mathfrak{d} + 2\mathfrak{g} - \mathfrak{a})\,\mathrm{tr}\,\mathfrak{M}\mathfrak{E}\,, \\ \varphi_2 &= (\mathfrak{d} + 2\mathfrak{g} - \mathfrak{a})\,\mathrm{tr}\,\mathfrak{E} + (\mathfrak{c} + \mathfrak{a} - 2\mathfrak{d} - 4\mathfrak{g})\,\mathrm{tr}\,\mathfrak{M}\mathfrak{E}\,, \\ \varphi_3 &= 2\mathfrak{g}\,, \end{aligned} \tag{20}$$

établissant ainsi le passage entre (17) et (19).

Références

1 Wang, C. C., A new representation theorem for isotropic functions. I, Arch. Rat. Mech. Analysis **35/36**, 166–197 (1970).
2 Wang, C. C., A new representation theorem for isotropc functions. II, Arch. Rat. Mech. Analysis **35/36**, 198–223 (1970).
3 Wang, C. C., Corrigendum, Arch. Rat. Mech. Analysis **43**, 392–395 (1971).
4 Boehler, J. P., Sur les formes invariantes dans le sous-groupe orthotrope de révolution des transformations orthogonales de la relation entre deux tenseurs symétriques du second ordre, ZAMM **55**, 609–611 (1975).
5 Spencer, A. J. M., Theory of invariants, Continuum Physics, ed. by C. Eringen, Academic Press 1971.

Eingereicht am 18. 7. 1975

Anschrift: Dr. J. P. Boehler, Institut de Mécanique, Université de Grenoble, B. P. 53 Centre de Tri, 38041 Grenoble-Cédex, FRANCE

ZAMM 55, 503–504 (1976)

J. Lensing

Verallgemeinertes Runge-Kutta-Verfahren

1. Einleitung

Die Erweiterung des Runge-Kutta-Verfahrens von [1] auf inhomogene Differentialgleichungssysteme erster Ordnung mit nichtkonstanten Koeffizienten gelingt durch eine besondere Anordnung der rechten Seite $\boldsymbol{r}_{(x)}$ in der Koeffizientenmatrix $\boldsymbol{A}_{(x)}$. Eine in $\boldsymbol{A}_{(x)}$ hinzugefügte Nullzeile gestattet dann gerade das zwanglose Aufaddieren der Steigerungswerte $\boldsymbol{k}_i$, sowohl beim Runge-Kuttaschen Mittelwert als auch vorher in den einzelnen $\boldsymbol{k}_i$-Schritten. Die Bezeichnungen werden in Anlehnung an [1] und [2] gewählt.

2. Besondere Anordnung

Von dem inhomogenen Differentialgleichungssystem erster Ordnung in der allgemeinen Schreibweise

$$\tilde{\boldsymbol{z}}'_{(x)} = \tilde{\boldsymbol{A}}_{(x)} \cdot \tilde{\boldsymbol{z}}_{(x)} + \boldsymbol{r}_{(x)} \tag{1}$$

wird die nichtkonstante Koeffizientenmatrix $\boldsymbol{A}_{(x)}$ um $\boldsymbol{r}_{(x)}$ und eine Nullzeile wie folgt erweitert:

$$\boldsymbol{A}_{(x)} := \begin{bmatrix} \tilde{\boldsymbol{A}}_{(x)} & \boldsymbol{r}_{(x)} \\ \boldsymbol{0}^\mathsf{T} & 0 \end{bmatrix}. \tag{2}$$

Auch der Zustandsvektor $\boldsymbol{z}_{(x)}$ muß um ein 1-Element erweitert werden, damit die Gleichung (1) wieder erfüllt ist:

$$\boldsymbol{z}_{(x)} := \begin{bmatrix} \tilde{\boldsymbol{z}}_{(x)} \\ 1 \end{bmatrix}. \tag{3}$$

Um nun die numerische Integration vornehmen zu können, wird der Integrationsbereich in Teilintervalle der Länge h_i unterteilt. Für jedes Teilintervall i $(i = 1, 2, 3, \ldots, N)$ wird eine Übertragungsmatrix $\boldsymbol{U}_i$ mit Hilfe des Runge-Kuttaschen Mittelwertes $\boldsymbol{k}_i$ ermittelt. Der Zustandsvektor $\boldsymbol{z}_i$ am Intervallende ergibt sich dann durch Matrizenmultiplikation der Übertragungsmatrix $\boldsymbol{U}_i$ mit dem Zustandsvektor $\boldsymbol{z}_{i-1}$ am Intervallanfang.

Für die Ermittlung der Übertragungsmatrix $\boldsymbol{U}_i$ werden der Einfachheit halber folgende Abkürzungen eingeführt:

$$\boldsymbol{A}_{i0} = \boldsymbol{A}_{(x_{i-1})}\,, \tag{4}$$

$$\boldsymbol{A}_{i1} = \boldsymbol{A}_{(x_{i-1}+h_i/2)}\,, \tag{5}$$

$$\boldsymbol{A}_{i2} = \boldsymbol{A}_{(x_{i-1}+h_i)}\,. \tag{6}$$

Bildet man nun den Runge-Kuttaschen Mittelwert

$$\boldsymbol{k}_i = \frac{1}{6}\,(\boldsymbol{k}_{i1} + 2\boldsymbol{k}_{i2} + 2\boldsymbol{k}_{i3} + \boldsymbol{k}_{i4})\,, \tag{7}$$

so erhält man mit den Zwischenwerten

$$\boldsymbol{k}_{i1} = h_i \boldsymbol{A}_{i0} \boldsymbol{z}_{i-1}\,, \tag{8}$$

$$\boldsymbol{k}_{i2} = (h_i \boldsymbol{A}_{i1} + (h_i^2/2)\,\boldsymbol{A}_{i1}\boldsymbol{A}_{i0})\,\boldsymbol{z}_{i-1}\,, \tag{9}$$

$$\boldsymbol{k}_{i3} = (h_i \boldsymbol{A}_{i1} + (h_i^2/2)\,\boldsymbol{A}_{i1}^2 + (h_i^3/4)\,\boldsymbol{A}_{i1}^2\boldsymbol{A}_{i0})\,\boldsymbol{z}_{i-1}\,, \tag{10}$$

$$\boldsymbol{k}_{i4} = (h_i \boldsymbol{A}_{i2} + h_i^2 \boldsymbol{A}_{i2}\boldsymbol{A}_{i1} + (h_i^3/2)\,\boldsymbol{A}_{i2}\boldsymbol{A}_{i1}^2 + (h_i^4/4)\,\boldsymbol{A}_{i2}\boldsymbol{A}_{i1}^2\boldsymbol{A}_{i0})\,\boldsymbol{z}_{i-1}\,, \tag{11}$$

die Übertragungsmatrix $\boldsymbol{U}_i$ direkt aus

$$\boldsymbol{z}_i = \boldsymbol{z}_{i-1} + \boldsymbol{k}_i \tag{12}$$

durch Einsetzen und Umordnen von (8—11) in (7):

$$\begin{aligned} \boldsymbol{z}_i = \Big[\boldsymbol{E} &+ h_i \left(\frac{1}{6} \cdot (\boldsymbol{A}_{i0} + 4\boldsymbol{A}_{i1} + \boldsymbol{A}_{i2})\right) + \\ &+ (h_i^2/2!) \cdot \left(\frac{1}{3}\,\boldsymbol{A}_{i1}\boldsymbol{A}_{i0} + \frac{1}{3}\,\boldsymbol{A}_{i1}^2 + \frac{1}{3}\,\boldsymbol{A}_{i2}\boldsymbol{A}_{i1}\right) + \\ &+ (h_i^3/3!) \cdot \left(\frac{1}{2}\,\boldsymbol{A}_{i1}^2\boldsymbol{A}_{i0} + \frac{1}{2}\,\boldsymbol{A}_{i2}\boldsymbol{A}_{i1}^2\right) + \\ &+ (h_i^4/4!) \cdot (\boldsymbol{A}_{i2}\boldsymbol{A}_{i1}^2\boldsymbol{A}_{i0})\Big]\,\boldsymbol{z}_{i-1}\,, \end{aligned} \tag{13}$$

oder kurz

$$\boldsymbol{z}_i = \boldsymbol{U}_i \boldsymbol{z}_{i-1}\,. \tag{14}$$

Mit den Abkürzungen für die runden Klammern $\boldsymbol{A}_i^{(1)}$ bis $\boldsymbol{A}_i^{(4)}$ folgt schließlich die *explizit formulierte Übertragungsmatrix für den Teilbereich i*

$$\boldsymbol{U}_i = \boldsymbol{E} + h_i \boldsymbol{A}_i^{(1)} + (h_i^2/2!)\,\boldsymbol{A}_i^{(2)} + (h_i^3/3!)\,\boldsymbol{A}_i^{(3)} + (h_i^4/4!)\,\boldsymbol{A}_i^{(4)}. \tag{15}$$

Die Gleichung (14) mit der Übertragungsmatrix (15) entspricht der numerischen Integration des inhomogenen Differentialgleichungs-Systems 1. Ordnung (1). Das Ergebnis ist in der Bezeichnungsweise identisch mit der homogenen Lösung [1], [2], hier ist lediglich nur die scheinbar triviale Erweiterung (2) zu beachten.

Eine Rückführung auf den homogenen Fall ist durch Nullsetzen von $\boldsymbol{r}_{(x)}$ und Streichen der entsprechenden Zeile und Spalte in $\boldsymbol{A}_{(x)}$, sowie Kürzen des Zustandsvektors $\boldsymbol{z}_{i-1}$ um das 1-Element direkt möglich.

Literatur

1 PESTEL, E.; LECKIE, F. A., Matrix Methods in Elastomechanics, New York 1963.
2 ZURMÜHL, R., Praktische Mathematik, Springer-Verlag, Berlin 1965.

Eingereicht am 24. 11. 1975

Anschrift: J. LENSING, D-33 Braunschweig, Lehrstuhl für Mechanik und Festigkeitslehre der TU Braunschweig. Postfach 7050.

ZAMM 56, 504 – 506 (1976)

G. AHMADI

Universal Stability of Micropolar Viscoelastic Fluid Motions

1. Introduction

The energy method of stability was first formulated by ORR [1]. The calculation procedure and the results associated with this older use of the energy method were summerized by BATEMAN, DRYDEN and MURNAGHAN [2]. These early results of the energy stability method gave such conservative estimate for the stability that the method was neglected for many years. The modern applications of the energy method dated from the work of SERRIN [3] who has developed a remarkable variational technique which gives results which are not too conservative. JOSEPH and his co-workers SHIR and CARMI in several papers [4—9] studied the universal stability of conducting — diffusing solutions. RAO [10] considered the stability of micropolar fluids. The stability criteria for COSSERAT fluids motions are established by SHAHINPOOR and AHMADI [11]. Recently, AHMADI and SHAHINPOOR [12, 13] studied the universal stability of magneto-micropolar and magneto-COSSERAT fluid motions. The sufficiency conditions for the stability of thermo-magneto-micropolar fluids motion are established by AHMADI [14]. The energy stability of a viscoelastic simple fluid heated from below is investigated by CARMI and SOKOLOV [15].

In the present work the theory of micropolar viscoelastic fluids as derived by ERINGEN [16] and discussed by ASKAR, CAKMAK and ARIMAN [17] is considered. The universal stability of the general unsteady motions of such fluid in a bounded domain is studied and the sufficiency conditions for the stability are obtained. Certain uniqueness theorems for the steady and unsteady motions of micropolar viscoelastic fluids are also established.

2. Theory of Micropolar Viscoelastic Fluids

The theories of micropolar media as derived by ERINGEN [16, 18, 19] are based on the following balance laws:

Mass

$$\frac{\partial\varrho}{\partial t} + (\varrho v_k)_{,k} = 0\,, \tag{2.1}$$

Linear momentum

$$t_{kl,k} + \varrho(f_l - \dot v_l) = 0\,, \tag{2.2}$$

Moment of momentum

$$m_{rk,r} + \varepsilon_{klr} t_{lr} + \varrho(l_k - j\dot\nu_k) = 0\,, \tag{2.3}$$

Energy

$$\varrho\dot\varepsilon = t_{kl} d_{lk} + \varepsilon_{klr} t_{kl}(\omega_r - \nu_r) + m_{kl}\nu_{l,k} + q_{k,k} + \varrho h \tag{2.4}$$

where

ϱ = mass density; t_{kl} = stress tensor; m_{kl} = couple stress tensor; j = microinertia; ε = internal energy; v_k = components of velocity vector $\boldsymbol v$; f_k = body force; l_k = body couple; ν_k = components of gyration vector $\boldsymbol\nu$; $\omega_k = \frac{1}{2}\varepsilon_{klm}v_{m,l}$ = angular velocity; q_k = heat flux vector; h = heat source; $d_{kl} = \frac{1}{2}(v_{k,l} + v_{l,k})$ = components of deformation rate tensor $\boldsymbol d$ (2.5)

Throughout the paper the regular cartesian tensor notation is employed with superposed dot standing for material derivative and an index following a comma denoting the partial differentiation. The properties of micropolar material are restricted by the entropy inequality

$$-\frac{\varrho}{\theta}(\dot\psi - \eta\dot\theta) + \frac{1}{\theta}t_{kl}d_{lk} + \frac{1}{\theta}\varepsilon_{klm}t_{kl}(\omega_m - \nu_m) + \frac{1}{\theta}m_{kl}\nu_{l,k} + \frac{q_k\theta_{,k}}{\theta^2} \geqq 0 \tag{2.6}$$

where θ is the temperature, η is the entropy and ψ is the free energy defined by

$$\psi = \varepsilon - \theta\eta\,. \tag{2.7}$$

Equations of balance must be supplemented by the constitutive equations. According to ERINGEN [16] the following linear constitutive equations are appropriate for viscoelastic fluids:

$$t_{kl} = -p\delta_{kl} + \int_{-\infty}^{t}\{\lambda(t-\zeta)\,d_{rr}(\zeta)\,\delta_{kl} + 2\mu(t-\zeta)\,d_{kl}(\zeta) + \varkappa(t-\zeta)\,\varepsilon_{klm}[\omega_m(\zeta) - \nu_m(\zeta)]\}\,d\zeta\,, \tag{2.8}$$

$$m_{kl} = \int_{-\infty}^{t}[\alpha(t-\zeta)\,\nu_{r,r}(\zeta)\,\delta_{kl} + \beta(t-\zeta)\,\nu_{k,l}(\zeta) + \gamma(t-\zeta)\,\nu_{lk}(\zeta)]\,d\zeta \tag{2.9}$$

where λ, μ, $\varkappa$, α, β, and γ are time dependent fading memory functions of the fluid and p is the pressure.

In the absence of body force and body couple and for an incompressible micropolar viscoelastic fluid the equations of motions are

$$\nabla\cdot\boldsymbol v = 0\,, \tag{2.10}$$

$$\varrho\dot{\boldsymbol v} = -\nabla p + \int_{-\infty}^{t}\{[\mu(t-\zeta) + \varkappa(t-\zeta)]\,\nabla^2\boldsymbol v(\zeta) + \varkappa(t-\zeta)\,\nabla\times\boldsymbol\nu(\zeta)\}\,d\zeta\,, \tag{2.11}$$

$$\varrho j\dot{\boldsymbol\nu} = \int_{-\infty}^{t}\{\varkappa(t-\zeta)\,[\nabla\times\boldsymbol v(\zeta) - 2\boldsymbol\nu(\zeta)] - \gamma(t-\zeta)\,\nabla\times\nabla\times\boldsymbol\nu(\zeta) + [\alpha(t-\zeta) + \beta(t-\zeta) + \gamma(t-\zeta)]\times\nabla(\nabla\cdot\boldsymbol\nu(\zeta))\}\,d\zeta\,.$$

The boundary conditions are assumed to be of superadherence form [10, 11]. That is

$$\boldsymbol v(\boldsymbol x, t) = \boldsymbol v_s(\boldsymbol x, t) \quad \text{on } S(t)\,, \tag{2.14}$$

$$\boldsymbol\nu(\boldsymbol x, t) = \boldsymbol\nu_s(\boldsymbol x, t) \quad \text{on } S(t)\,. \tag{2.15}$$

Employing the maximum length in the domain L_0, the maximum reference velocity V_0 and using the same notation for the dimensionless quantities, equations of motion becomes

$$\nabla\cdot\boldsymbol v = 0\,, \tag{2.16}$$

$$\dot{\boldsymbol v} = -\nabla p + \left(\frac{M+K}{Re}\right)*\nabla^2\boldsymbol v + \frac{K}{Re}*\nabla\times\boldsymbol\nu\,, \tag{2.17}$$

$$j\dot{\boldsymbol\nu} = \frac{K}{Re}*(\nabla\times\boldsymbol v - \zeta\boldsymbol\nu) - \frac{C}{Re}*\nabla\times\nabla\times\boldsymbol\nu + \frac{D}{Re}*\nabla(\nabla\boldsymbol\nu) \tag{2.18}$$

where the asterisk stands for the convolution product and

$$Re = \frac{\varrho V_0 L_0}{\mu_0}\,, \qquad \mu_0 = \int_0^\infty \mu(\zeta)\,d\zeta\,,$$

$$M(\zeta) = \mu(\zeta)/\mu_0\,, \qquad K(\zeta) = \varkappa(\zeta)/\mu_0\,,$$

$$C(\zeta) = \gamma(\zeta)/\mu_0 L_0^2\,, \qquad D(\zeta) = (\alpha(\zeta) + \beta(\zeta) + \gamma(\zeta))/\mu_0 L_0^2\,. \tag{2.19}$$

These, together with boundary conditions (2.19) and (2.16) are the basic equation of motion of a micropolar viscoelastic fluid in dimensionless form.

3. Universal Stability Analysis

Suppose we are interested in the stability of a fluid motion $(\boldsymbol{v}, \boldsymbol{\nu}, p)$ which satisfied the differential equations (2.16 to 2.18) and the boundary conditions (2.14) and (2.15). Now we assume that the basic flow $(\boldsymbol{v}, \boldsymbol{\nu}, p)$ is disturbed to $(\boldsymbol{v} + \boldsymbol{u}, \boldsymbol{\nu} + \boldsymbol{Q}, p + \pi)$ which must also satisfy the differential equation and boundary conditions. In order to determine whether or not the disturbed flow approaches the basic flow asymptotically in the mean as $t \to \infty$, we employ a positive definite LYAPUNOV's measure in the form of the energy of the difference motion with its velocity $\boldsymbol{u}$, microrotation $\boldsymbol{Q}$ and pressure π. The energy of the difference motion is given by

$$T = T_1 + T_2 \tag{3.1}$$

where[1])

$$T_1 = \frac{1}{2}\int u^2 \qquad T_2 = \frac{1}{2}\int jQ^2 . \tag{3.2}$$

The equations for the dynamics of difference motion $(\boldsymbol{u}, \boldsymbol{Q}, \pi)$ can be easily obtained by subtracting equations (2.16—2.18) for the basic motion from the same equations written for the disturbed motion. These are

$$\nabla \cdot \boldsymbol{u} = 0 , \tag{3.3}$$

$$\frac{\partial \boldsymbol{u}}{\partial t} + \boldsymbol{u} \cdot \nabla \boldsymbol{v} + \boldsymbol{v} \cdot \nabla \boldsymbol{u} + \boldsymbol{u} \cdot \nabla \boldsymbol{u} = -\nabla \pi + \left(\frac{M+K}{Re}\right) * \nabla^2 \boldsymbol{u} + \frac{K}{Re} * \nabla \times \boldsymbol{Q} , \tag{3.4}$$

$$j\left(\frac{\partial \boldsymbol{Q}}{\partial t} + \boldsymbol{u} \cdot \nabla \boldsymbol{\nu} + \boldsymbol{v} \cdot \nabla \boldsymbol{Q} + \boldsymbol{u} \cdot \nabla \boldsymbol{Q}\right) = \frac{K}{Re} * (\nabla \times \boldsymbol{u} - 2\boldsymbol{Q}) - \frac{C}{Re} * \nabla \times \nabla \times \boldsymbol{Q} + \frac{D}{Re} * \nabla(\nabla \cdot \boldsymbol{Q}) . \tag{3.5}$$

Since both motions satisfy the same boundary conditions, then

$$\boldsymbol{u} = \boldsymbol{Q} = 0 \qquad \text{on} \qquad S(t) . \tag{3.6}$$

Dotting equation (3.4) by $\boldsymbol{u}$, integrating over the volume $V(t)$ and making use of (3.3) and divergence theorem we find

$$\frac{\mathrm{d}}{\mathrm{d}t}\int \frac{u^2}{2} = \int\left[-\boldsymbol{u} \cdot \nabla \boldsymbol{v} \cdot \boldsymbol{u} - \frac{1}{Re}(M+K) * (\nabla \times \boldsymbol{u}) \cdot (\nabla \times \boldsymbol{u}) + \frac{1}{Re} K * \boldsymbol{Q} - \nabla \times u\right] + \int_s \left[-\boldsymbol{v}\frac{u^2}{2} - \boldsymbol{u}\frac{u^2}{2} - \boldsymbol{u}\pi + \frac{1}{Re} K * \boldsymbol{Q} \times \boldsymbol{u} - \frac{1}{Re}(M+K) * (\nabla \times \boldsymbol{u}) \times \boldsymbol{u}\right] \cdot \mathrm{d}\boldsymbol{s} . \tag{3.7}$$

Similarly, dotting equation (3.5) by $\boldsymbol{Q}$ yields

$$\frac{\mathrm{d}}{\mathrm{d}t}\int \frac{jQ^2}{2} = \int\left[-\boldsymbol{u} \cdot \nabla \boldsymbol{\nu} \cdot \boldsymbol{Q} - \frac{1}{Re} C * (\nabla \times \boldsymbol{Q}) \cdot (\nabla \cdot \boldsymbol{Q}) - \frac{1}{Re} D * (\nabla \cdot \boldsymbol{Q})(\nabla \cdot \boldsymbol{Q}) + \frac{K}{Re} * (\nabla \times \boldsymbol{u} - 2\boldsymbol{Q}) \cdot \boldsymbol{Q}\right] + \int_s \left[-\boldsymbol{v}\frac{Q^2}{2} - \boldsymbol{u}\frac{Q^2}{2} - \frac{1}{Re} C * (\nabla \times \boldsymbol{Q}) \times \boldsymbol{Q} + \frac{1}{Re} D * (\nabla \cdot \boldsymbol{Q})\,\boldsymbol{Q}\right] \cdot \mathrm{d}\boldsymbol{s} . \tag{3.8}$$

[1]) The conventional volume infinitesimal is omitted consistently in the integrals which are extended over the volume $v(t)$.

Note that the surface integral in (3.7) and (3.8) vanish because of the boundary conditions (3.6).

It is then found that

$$\frac{\mathrm{d}T}{\mathrm{d}t} = -\int \boldsymbol{u} \cdot \boldsymbol{d} \cdot \boldsymbol{u} - \int\Big[\boldsymbol{u} \cdot \nabla \boldsymbol{\nu} \cdot \boldsymbol{Q} - \frac{1}{Re} K * \boldsymbol{Q} \cdot \boldsymbol{Q} + K * (\boldsymbol{Q} - \nabla \times \boldsymbol{u}) \cdot (\boldsymbol{Q} - \nabla \times \boldsymbol{u}) + M * (\nabla \times \boldsymbol{u}) \cdot (\nabla \times \boldsymbol{u}) + C * (\nabla \times \boldsymbol{Q}) \cdot \nabla \times \boldsymbol{Q} + D * (\nabla \cdot \boldsymbol{Q})(\nabla \cdot \boldsymbol{Q})\Big] . \tag{3.9}$$

If M, K, C, and D are assumed to be constants and the convolution integral are replaced by simple multiplication, then (3.9) becomes identical with the results of RAO [10], AHMADI and SHAHINPOOR [12] and AHMADI [14].

Equation (3.9) gives the dynamics of energy of the difference motion. For the basic motion to be stable it is sufficient that the energy of the difference motion T decay to zero as time increases. Hence, the condition $\mathrm{d}T/\mathrm{d}t < 0$ for all t insures that $T \to 0$ as $t \to \infty$ and therefore is a sufficient condition for asymptotic stability in the mean.

Equation (3.9) may be rewritten as

$$\frac{\mathrm{d}T}{\mathrm{d}t} = \frac{J}{Re}\left[-1 + Re\left(\frac{-I}{J}\right)\right] \leqq \frac{J}{Re}\left[-1 + \frac{Re}{R}\right] \tag{3.10}$$

where

$$\frac{1}{R} = \max_{\boldsymbol{u}, \boldsymbol{Q}} (-I/J) \tag{3.11}$$

with

$$J = \int [K * \boldsymbol{Q} \cdot \boldsymbol{Q} + K * (\boldsymbol{Q} - \nabla \times \boldsymbol{u}) \cdot (\boldsymbol{Q} - \nabla \times \boldsymbol{u}) + M * (\nabla \times \boldsymbol{u}) \cdot \nabla \times \boldsymbol{u} + C * (\nabla \times \boldsymbol{Q}) \cdot (\nabla \times \boldsymbol{Q}) + D * (\nabla \cdot \boldsymbol{Q})(\nabla \cdot \boldsymbol{Q})] , \tag{3.12}$$

$$I = \int [\boldsymbol{u} \cdot \boldsymbol{d} \cdot \boldsymbol{u} + \boldsymbol{u} \cdot \nabla \boldsymbol{\nu} \cdot \boldsymbol{Q}] \tag{3.13}$$

for given M, K, C, and D, and the constraint

$$\nabla \cdot \boldsymbol{u} = 0 . \tag{3.14}$$

The extremum problem (3.11)—(3.14) is equivalent to the following variational description

$$\delta\left[I(\boldsymbol{u}, \boldsymbol{\nu}) + \frac{1}{R} J(\boldsymbol{u}, \boldsymbol{\nu}) - 2\int p\nabla \cdot \boldsymbol{u}\right] = 0 . \tag{3.15}$$

The corresponding EULER-LAGRANGE equations are

$$\boldsymbol{u} \cdot \boldsymbol{d} + \frac{1}{2} j \nabla \boldsymbol{\nu} \cdot \boldsymbol{Q} = -\nabla p + \frac{(M+K)}{R} * \nabla^2 \boldsymbol{u} + \frac{K}{R} * \nabla \times \boldsymbol{Q} , \tag{3.16}$$

$$\frac{1}{2} j\boldsymbol{u} \cdot \nabla \boldsymbol{\nu} = \frac{K}{R} * (\nabla \times \boldsymbol{u} - \boldsymbol{Q}) + \frac{C}{R} * \nabla^2 \boldsymbol{Q} + \frac{(D-C)}{R} * \nabla\nabla \cdot \boldsymbol{Q} , \tag{3.17}$$

$$\nabla \cdot \boldsymbol{u} = 0 . \tag{3.18}$$

The least positive eigenvalue R of (3.16—3.18) for given M, K, C, and D gives the limit stability point where for

$$Re < R \tag{3.19}$$

$\mathrm{d}T/\mathrm{d}t$ from equation (3.10) remains negative and hence the universal stability of the basic flow is ensured. Therefore the following criteria may be deduced:

Theorem 1 — *If $M(t)$, $K(t)$, $C(t)$, $D(t)$, $\boldsymbol{d}$, $\nabla \boldsymbol{\nu}$ and Re for a micropolar viscoelastic fluid flow in a bounded arbitrary time dependent domain $v(t)$ of space satisfy the condition*

$$Re < R \tag{3.20}$$

where R is the least positive eigenvalue of (3.16—3.18), then the energy of the difference motion T approaches zero asymptotically in the mean as $t \to \infty$ and the basic flow $(\boldsymbol{v}, \boldsymbol{\nu})$ is stable.

From Theorem 1 the following corollaries can be deduced:
Corollary 1. (Uniqueness of unsteady micropolar viscoelastic fluid motions.) *If* ($\boldsymbol{v}$, $\boldsymbol{\nu}$, p) *and* ($\boldsymbol{v} + \boldsymbol{u}$, $\boldsymbol{\nu} + \boldsymbol{Q}$, $p + \pi$) *are two unsteady micropolar viscoelastic fluid motions in a bounded domain* $v(t)$ *of space having the same velocity, and microrotation at* $t = 0$ *and on the boundary of* $v(t)$, *namely*, $s(t)$, *then they must be identical.*

The proof of the Corollary is easily seen by noting that the energy of difference motion satisfies equation (3.10) and hence must be a decreasing function of time. But since at $t = 0$, $\boldsymbol{u} = \boldsymbol{Q} = 0$, we find that $T(t) = 0$ for all $t \geqq 0$ and hence the two motions are identical.

Corollary 2. (Uniquiness of steady micropolar viscoelastic fluid motions.) *If* ($\boldsymbol{v}$, $\boldsymbol{\nu}$, p)*and* ($\boldsymbol{v} + \boldsymbol{u}$, $\boldsymbol{\nu} + \boldsymbol{Q}$, $p + \pi$) *are two steady motions of* v *micropolar viscoelastic fluid in a bounded domain* $v(t)$ *of space having the same velocity and microrotation on the boundary of* v, *namely* s, *then they must be identical.*

The proof of the corollary is easily seen by studying the energy of the difference motion which must be a constant since both motions are steady. But T must satisfy equation (3.10) and hence is a decreasing function of t. This is only possible if $T = 0$ for all $t \geqq 0$ implying that the motions are identical.

Acknowledgments

The partial financial support of the Plasma Physics Division of the Department of Physics of the University of Saskatchewan is gratefully acknowledged.

References

1 ORR, W. McF., Proc. Roy. Irish. Acad. (A) **27**, 69 (1907).
2 BATEMAN, H., DRYDEN, H., and MURNAGHAN, F. D., Hydrodynamics, New York, Dover, (1932).
3 SERRIN, J., Arch. Rat. Mech. Anal. **3**; 1 (1959).
4 JOSEPH, D. D., Arch. Rat. Mech. Anal. **22**, 163 (1966).
5 SHIR, C. C. and JOSEPH, D. D., Arch. Rat. Mech. Anal. **30**, 38 (1968).
6 JOSEPH, D. D. and CARMI, S. J., J. Fluid Mech. **26**, 769 (1966).
7 JOSEPH, D. D. and SHIR, C. C., J. Fluid Mech. **26**, 753 (1966).
8 JOSEPH, D. D., Arch. Rat. Mech. Anal. **36**, 285 (1970).
9 JOSEPH, D. D., Arch. Rat. Mech. Anal. **44**, 1 (1971).
10 RAO, S. K. L., Int. J. Engng. Sci. **8**, 753 (1970).
11 SHAHINPOOR, M. and AHMADI, G., Arch. Rat. Mech. Anal. **47**, 183 (1972).
12 AHMADI, G. and SHAHINPOOR, M., Int. J. Engng. Sci. **12**, 657 (1974).
13 AHMADI, G. and SHAHINPOOR, M., Rend. Sci. A, **107**, 343 (1973).
14 AHMADI, G., Int. J. Engng. Sci. (to appear).
15 CARMI, S. and SOKOLOV, M., Phys. Fluids **17**, 544 (1974).
16 ERINGEN, A. C., Int. J. Engng. Sci. **5**, 191 (1967).
17 ASKAR, A., CAKMAK, A. S., and ARIMAN, T., Int. J. Engng. Sci. **6**, 283 (1968).
18 ERINGEN, A. C., Proc. XI Intern. Congress of Appl. Mech., Springer-Verlag (1965).
19 ERINGEN, A. C., J. Math. Mech. **16**, 1 (1966).

Eingereicht am 30. 4. 1975

Anschrift: Dr. G. AHMADI, Department of Mechanical Engineering, Pahlavi University Shiraz, IRAN.

BUCHBESPRECHUNGEN

Distéfano, N., Nonlinear Processes in Engineering Mechanics, 366 S., New York. Academic Press. 1974. \$ 26.50. £ 12.70.

Die mathematische Behandlung technischer Probleme hat mit dem Auftreten elektronischer Rechenautomaten neue, wesentliche Impulse erhalten. Die Leistungsfähigkeit der Computer führte im verstärkten Maße zu numerischen Untersuchungen und Lösungsmöglichkeiten auch von Aufgaben, die früher nicht oder nur in groben Näherungen behandelt werden konnten. Dies hatte zur Folge, daß viele neue mathematische Ideen entwickelt und mit bekannten Methoden zu wirkungsvollen Berechnungsverfahren kombiniert wurden. Gleichzeitig ist ein allgemeineres Herangehen an die Modellierung physikalischer Phänomene zu verzeichnen. Dieser neuen Methodologie ist die vorliegende Monographie gewidmet. Sie gibt eine Einführung in Methoden und Techniken für das Studium nichtlinearer Prozesse, die in der Beschreibung technischer Vorgänge in vielfältiger Weise auftreten. Insbesondere behandelt der Autor dynamische technische Prozesse unter Benutzung moderner mathematischer Methoden, wie der invarianten Einbettung, der dynamischen Programmierung, der Quasilinearisierung, der Methode der finiten Elemente, der Systemidentifikation, der Optimierung und der Kontrolltheorie.

Das Buch ist in zwölf Kapitel gegliedert. Das erste Kapitel ist der Einführung und der numerischen Handhabung der später häufig benutzten Methode der invarianten Einbettung gewidmet. Im zweiten Kapitel wird ein kurzer Abriß über Systeme von gewöhnlichen Differentialgleichungen, über Anfangs- und Randwertprobleme und deren numerische Lösung angegeben. Das dritte Kapitel beschäftigt sich mit balkenähnlichen Strukturen und Methoden zu deren Berechnung, wie invariante Einbettung, Störungsmethode, RICCATI-Transformation, Differenzenmethode, Übertragungsmatrizenmethode. Im vierten und fünften Kapitel werden spezielle Probleme behandelt, die auf partielle Differentialgleichungen führen. Zur Lösung der Poissonschen Gleichung sind Diskretisierungsmethoden (Methode der finiten Elemente, Differenzenmethode) und zur Behandlung der rechteckigen elastischen Platte Semidiskretisierungsmethoden beschrieben. Weiterhin werden analytische Methoden zur Behandlung partieller Differentialgleichungen angeführt (RICCATI-Transformation, invariante Einbettung). Im sechsten Kapitel wird auf einige grundlegende Ideen, Techniken und numerische Lösungswege der dynamischen Programmierung eingegangen und auf die Verwendbarkeit dieser Methoden zur Behandlung von Problemen der Festkörpermechanik hingewiesen. Die Methode der Quasilinearisation, insbesondere die Methode der sukzessiven Approximation, ist im siebenten Kapitel entwickelt. Die Verwendung vorher entwickelter mathematischer Methoden auf einige Probleme der Festkörpermechanik ist im achten Kapitel gezeigt; es werden elastische Gebinde, Balken, Sandwichbalken und Rotationsschalen behandelt. Methoden der Kontrolltheorie und deren Verwendung bei technischen Problemen, insbesondere in der Festkörpermechanik, sind Gegenstand des neunten Kapitels. Die weiteren Kapitel behandeln spezielle Aufgabenstellungen der Festkörpermechanik. Im zehnten und elften Kapitel wird die Anwendung der Methoden der dynamischen Programmierung und der invarianten Einbettung zur Berechnung optimaler Balken- und Scheibenkonstruktionen beschrieben. Im abschließenden zwölften Kapitel werden ausführlich Identifikationsprobleme in der Festkörpermechanik sowie analytische und numerische Lösungsverfahren angegeben.

Die behandelten, in verständlicher Form dargebotenen Methoden sind an Beispielen erläutert und durch eine Vielzahl von Übungsaufgaben ergänzt. Jedes Kapitel endet mit einer umfangreichen Literaturübersicht, in der auch auf weiterführende Probleme hingewiesen wird.

Berlin H. FRIEDRICH

G. W. Bluman / J. D. Cole, Similarity Methods for Differential Equations. 43 figs. IX, 332 pages. Berlin-Heidelberg-New York. Springer-Verlag. 1974. Soft cover DM 23,30; US \$ 9.60.

Das Buch ist einer systematischen Integrationstheorie von gewöhnlichen und partiellen Differentialgleichungen gewidmet, die auf der Invarianz dieser Gleichungen unter stetigen Transformationsgruppen (LIEsche Gruppen) beruht. Das Wesen dieser Methode, die im vergangenen Jahrhundert von SOPHUS LIE begründet wurde, besteht in folgendem: Falls eine stetige Transformationsgruppe bekannt ist, die die zu betrachtende Differentialgleichung invariant läßt, so kann bei einer gewöhnlichen Differentialgleichung die Ordnung der

Gleichung erniedrigt werden (im Falle einer Gleichung erster Ordnung läßt sich folglich die Lösung mittels Quadratur bestimmen), während bei einer partiellen Differentialgleichung zumindest die Anzahl der unabhängigen Variablen reduziert werden kann (in günstigen Fällen läßt sich eine Zurückführung auf gewöhnliche Differentialgleichungen erreichen). Der Nachteil dieser Methode besteht darin, daß es kein systematisches Verfahren weder für das Auffinden der Transformationsgruppen noch zum Nachweis ihrer Existenz oder Nichtexistenz gibt.

Das vorliegende Buch besteht in natürlicher Weise aus zwei Teilen. Teil 1 (142 Seiten) ist den gewöhnlichen Differentialgleichungen gewidmet, Teil 2 (182 Seiten) den partiellen. Teil 1 enthält neben einer anschaulichen Darlegung des Wesens der LIEschen Integrationsmethode die Bestimmung der Gleichungen erster und zweiter Ordnung, die unter einer gegebenen stetigen Transformationsgruppe invariant sind.

Als Anwendungen dieser Methode werden neben den bekannten Beispielen der THOMAS-FERMI-Gleichung sowie der BLASIUS-Gleichung auch die Deformation einer Membran unter Einwirkung einer axialsymmetrischen Belastung behandelt.

Im Teil 2 werden neben den allgemeinen Darlegungen über die Anwendung von stetigen Transformationsgruppen auf partielle Differentialgleichungen die Beziehungen dieser Integrationstheorie zur Theorie der Ähnlichkeitslösungen und der Dimensionsanalyse behandelt. Die Konstruktion der allgemeinen Ähnlichkeitslösung der eindimensionalen Wärmeleitungsgleichung stellt das prinzipielle Beispiel für die Anwendung stetiger Transformationsgruppen auf partielle Differentialgleichungen dar, andere Beispiele sind die (eindimensionale) axialsymmetrische Wellengleichung, die (eindimensionale) FOKKER-PLANCK-Gleichung sowie einige spezielle Versionen nichtlinearer Wärmeleitungs- und Wellengleichungen.

Berlin K. R. SCHNEIDER

Beck, A., Continuous Flows in the Plane. 47 figs. XI, 462 pages. Berlin-Heidelberg-New York. Springer-Verlag. 1974. Cloth DM 114,– ; US $ 46.80.

Hervorgegangen aus der qualitativen Theorie gewöhnlicher Differentialgleichungen ist die topologische Dynamik heute eine eigene Theorie, die eher der allgemeinen oder geometrischen Topologie als der Analysis zuzurechnen ist. Ihr Forschungsobjekt sind die stetigen Operationen der Gruppe $\mathbb{R}$ der reellen Zahlen (oder manchmal auch der Gruppe der ganzen Zahlen) auf einem topologischen Raum T, d. h. die auch als Flüsse bezeichneten Abbildungen $\varphi\colon T \times \mathbb{R} \to T$ mit der Eigenschaft $\varphi(p, 0) = p$, $\varphi(p, s + t) = \varphi(\varphi(p, s), t)$. Führen schon die Differentialgleichungen selbst (bei denen ja noch gewisse Differenzierbarkeitsvoraussetzungen erfüllt sind) zu einer so reichen Fülle von geometrischen Erscheinungen (erinnert sei nur an die Homoklininizität), so gilt dies erst recht für die Verallgemeinerung auf den topologischen Fall, und es sind Beschränkungen notwendig, um zu interessanten Ergebnissen zu gelangen. Eine solche Einschränkung, die zu einer reizvollen Theorie führt, erhält man, wenn man nicht beliebige topologische Räume T sondern nur die Ebene $\mathbb{R}^2$ betrachtet. Es ist die Gültigkeit des JORDANschen Kurvensatzes, die hier eine gewisse Ordnung in die bei höherdimensionalen Räumen unentwirrbare Vielfalt von Trajektorien bringt, obwohl man auch hier noch weit entfernt ist von einem Überblick über alle Phasenprofile (das sind die Aufteilungen von $\mathbb{R}^2$ in die einzelnen Trajektorien).

Im vorliegenden Buch wird eine Fülle von Ergebnissen geboten, die die Struktur von solchen Flüssen auf $\mathbb{R}^2$ wenigstens teilweise beschreiben. Dabei fließen neben den „klassischen" Resultaten etwa von W. KAPLAN und L. MARKUS viele vom Autor und seinen Mitarbeitern erzielte Ergebnisse ein. Insbesondere werden auf etwa 150 Seiten zwei Dissertationen von J. W. LEWIN bzw. M. H. LEWIN über Reparametrisierung bzw. algebraische Kombination von Flüssen auf allgemeinen Räumen T wiedergegeben. Um nur einen oberflächlichen Einblick in die Methoden zu geben, sei angedeutet, daß eine Möglichkeit, allgemeine Flüsse auf R^2 zu beschreiben, darin besteht, die Ebene in Zellen zu zerlegen, in denen sich der Fluß einfach verhält, und zu sehen, wie sich diese Zellen zusammenfügen. Im allgemeinen sind die Ergebnisse jedoch nicht einfach zu formulieren, und es erfordert einige Hingabe an diese recht speziellen aber geometrisch reizvollen Probleme, um die Bedeutung der komplizierten Definitionen und Aussagen zu erfassen.

Berlin H.-G. BOTHE

Müller, H. H./Magnus, K., Übungen zur Technischen Mechanik, 292 S., Stuttgart. B. G. Teubner. 1974. DM 24,–.

Das Verständnis der Technischen Mechanik hängt im starken Maße von der Fertigkeit im Lösen spezieller Aufgaben ab. Die meisten Lücken offenbaren sich erst beim selbständigen Üben, das um so effektiver ist, je sicherer Aufgabenstellung, Lösungsansatz und Lösung den Studenten ansprechen.

Das vorliegende Buch gibt in 136 Aufgaben einen Querschnitt durch die Grundlagen des Fachgebietes, der nicht nur repräsentativ, sondern auf Grund der Auswahl auch interessant ist. Nach einem kurzen Abschnitt über Vektoren werden 33 Aufgaben aus der Stereo-Statik behandelt. Die Elasto-Statik nimmt den gleichen Raum ein, es wird auf Spannungsermittlungen und Verformungsberechnungen eingegangen.

Ein kleinerer Abschnitt behandelt Aufgaben der Fluid-Statik. Auch die Kinematik ist in wenigen Beispielen vertreten. Diese wenden jedoch die Vektordarstellung konsequent an und scheuen auch nicht vor schwieriger Fragestellung, wie sie beispielsweise bei Kardangelenken vorliegt.

Der größte Abschnitt behandelt Aufgaben der Kinetik. Hierbei geht man konsequent vom Dynamischen Grundgesetz aus, das als Impulssatz bezeichnet wird. Die Lagrangeschen Gleichungen 2. Art werden angeführt, das D'ALEMBERTsche Prinzip findet keine Erwähnung.

Alle Aufgaben enthalten ausführlichen Lösungsweg und das Ergebnis, das in den meisten Fällen durch Zahlenwerte verdeutlicht wird. Neben verschiedenen historischen Aufgaben stehen eine große Zahl von Novitäten, die nicht nur für Studenten von Interesse sind. Bestimmt hat dieses schöne Buch schon viele Freunde gefunden.

Dresden F. HOLZWEISSIG

Martos, B., Nonlinear Programming Theory and Methods, 280 S., Budapest. Akadémiai Kiadó. 1975. Ft 180,–.

Innerhalb der mathematischen Optimierung wird seit vielen Jahren am Aufbau einer abgerundeten einheitlichen Theorie der nichtlinearen (speziell konvexen) Optimierung gearbeitet. Zwei Hauptlinien lassen sich feststellen: in Erweiterung des klassischen Besitzstandes werden die Untersuchungen einmal auf allgemeine lineare Räume ausgedehnt und zum anderen der Begriff der Konvexität so weit wie möglich durch schwächere Begriffe ersetzt. Das vorliegende Buch ordnet sich in die zweite Gruppe ein. Es liegt in einer Reihe mit den Büchern von C. BERGE / A. GHOUILA-HOURI ("Programmes, jeux et reseanx de transport", 1962) und von O. L. MANGASARIAN ("Nonlinear Programming", 1969). Der Einfluß der zuletzt genannten Darstellung ist besonders deutlich; eine Reihe von Teilthemen und Grundideen sind direkt übernommen. Im Gegensatz zu MANGASARIAN wird aber mit Erfolg versucht, durch eine konsequente Einbeziehung der Quasikonvexität bzw. Pseudokonvexität von Anfang an die Theorie geschlossen und abgerundet darzustellen (Teil I). Der Verfasser kann für größere Abschnitte der konvexen Optimierung notwendige und hinreichende Aussagen erreichen. Eine Reihe von Beweisen sind eleganter geworden (auf Grund vieler eigener Ergebnisse und der Einbeziegung neuer Publikationen) und die Gesamtdarstellung kürzer und übersichtlicher. Zahlreiche Gegenbeispiele verdeutlichen die Grenzen des eingesetzten Apparates, allerdings nicht durchgängig

Der theoretische Teil I umfaßt neun Kapitel: Mengen und Funktionen, Das nichtlineare Optimierungsproblem, (Quasi-) konvexe Funktionen und verwandte Konzepte, Zulässige Mengen (hier finden sich u. a. Trennungsaussagen und Alternativsätze), Minima, LAGRANGEsche Sattelpunkte, Differenzierbare Funktionen und Pseudokonvexität, Stationäre KUHN-TUCKER-LAGRANGE-Punkte und Dualität, Anhang: Konvexitätseigenschaften quadratischer Formen (den Verallgemeinerungen des Konvexitätsbegriffs wird die sog. Subdefinitheit von Matrizen gegenübergestellt). Es bleibt anzumerken, daß in der Darstellung — wie aus der o. g. Konzeption des Buches bereits hervorgeht — der Dualität kein zentraler Platz zugewiesen wird; duale Aufgaben werden über die verallgemeinerte LAGRANGEfunktion definiert und einige ausgewählte Ergebnisse als Abschluß des theoretischen Teils vorgeführt. Teil II bringt einige Methoden zur Lösung nichtlinearer Optimierungsprobleme: nach einer kurzen Verständigung über den Grundansatz der Simplexverfahren wird im Kapitel 11 deren Erweiterung auf Aufgaben mit pseudomonotoner und streng quasimonotoner Zielfunktion behandelt (mit dem wichtigen Fall der Quotientenoptimierung); Kapitel 12 bringt zwei Verfahren der vollständigen Beschreibung (das von MANAS und NEDOMA sowie das von UZAWA); für quadratische Aufgaben wird WOLFES Methode in seiner kurzen wie langen Form angegeben und letztere auch im Fall quasikonvexer Zielfunktionen angewendet, als Beispiel für Gradientenverfahren bringt der Autor die heute wieder stark beachtete Methode der bedingten Gradienten von FRANK und WOLFE; das Schlußkapitel 15 beinhaltet Schnittverfahren (KELLEY, VEINOTT). Allen beschriebenen Verfahren sind Konvergenzaussagen und -beweise beigefügt sowie kleine Beispiele; es fehlen jedoch Mitteilungen über numerische Erfahrungen, Güteaussagen über die Algorithmen u. ä. Die Auswahl der angegebenen Verfahren ist durch die Konzeption des ersten Teils des Buches bedingt; Teil II trägt daher eher Beispielcharakter und kann kaum als repräsentativ für den heutigen Stand der nichtlinearen Optimierung angesehen werden (so ist z. B. der Verzicht auf Straffunktionsverfahren u. ä. von der Anlage des Buches her einzusehen, aber nicht der Verzicht auf die Zentrenmethoden).

Das Buch ist flüssig und leserfreundlich geschrieben, alle Schritte werden ausführlich motiviert, die hergeleiteten Eigenschaften auf ihre Überprüfbarkeit untersucht (wenn auch oft nur verbal) und die Grenzen der Aussagen abgesteckt. Der Leser erhält auf einem Teilgebiet der nichtlinearen Optimierung eine in sich geschlossene Darstellung des Stoffes, Ein gründliches Literaturverzeichnis sowie die bibliographischen Anmerkungen am Ende jedes Kapitels geben auch dem weniger erfahrenen Leser Einblick in den Stand und die bisherige Entwicklung der jeweils behandelten Materie.

Berlin K. LOMMATZSCH

Baker, A., Transcendental Number Theory, 148 S., Cambridge. Cambridge University Press. 1975. £ 4.90.

Dieses Buch bietet dem Leser moderne Mathematik aus erster Hand. Der Autor ist in höchstem Maße kompetent für die Theorie der transzendenten Zahlen, da seine eigenen Beiträge (seit Anfang der 60er Jahre) zu den bahnbrechenden Leistungen auf diesem Gebiet gehören und auf dem Internationalen Mathematiker-Kongreß Nizza 1970 durch die Verleihung der Fields-Medaille (der höchsten Auszeichnung durch die Internationale Mathematiker-Union) gewürdigt wurden. — Eine komplexe Zahl α heißt *algebraisch*, wenn sie einer Gleichung $a_0\alpha^n + a_1\alpha^{n-1} + \cdots + a_n = 0$ mit ganzzahligen (!) Koeffizienten $a_0, a_1, \ldots, a_n$ $a_0 \neq 0$, genügt. (Wenn man den Exponenten n minimal wählt, dann ist diese definierende Gleichung bis auf einen konstanten Faktor durch α bestimmt, und dieses n heißt dann der Grad von α.) Eine komplexe Zahl heißt *transzendent*, wenn sie nicht algebraisch ist. Die Menge der algebraischen Zahlen bildet einen Teilkörper des Körpers der komplexen Zahlen und ist abzählbar. Daher sind „fast alle“ komplexen Zahlen transzendent. So sind z. B. die bekannten Zahlen π, e, $e^{\pi} = (-1)^{-i}$ und $\cos\alpha$, $\sin\alpha$, $\tan\alpha$ für fast alle algebraischen Argumente α transzendent. Im Beweis dieser Tatsache kulminierte die Theorie der transzendenten Zahlen im 19. Jahrhundert. Dagegen ist bis heute unbekannt, ob z. B. $e + \pi$ transzendent ist. Der Autor versteht es, mit vorbildlicher Klarheit auf knapp 130 Seiten (ergänzt durch ein Literaturverzeichnis von 16 Seiten) die grundlegenden Ergebnisse mit Beweisen und aktuelle Probleme des Gebiets darzulegen. Die Überschriften der 12 Kapitel lauten: 1. The origins; 2. Linear forms in logarithms; 3. Lower bounds for linear forms; 4. Diophantine equations; 5. Class numbers of imaginary quadratic fields; 6. Elliptic functions; 7. Rational approximations to algebraic numbers; 8. MAHLERS classification; 9. Metrical theory; 10. The exponential function; 11. The SIEGEL-SHIDLOVSKY theorems; 12. Algebraic independence. Die Gegenstände der Kapitel 1—3 und 7—12 lassen sich wie viele Probleme über transzendente Zahlen mit verhältnismäßig wenigen Begriffen formulieren, gehören aber z. T. zu den schwierigsten Fragen und haben eine lange Vorgeschichte. Kap. 2 enthält das zentrale Ergebnis des Verfassers: Wenn $\alpha_1, \ldots, \alpha_n$ nichtverschwindende algebraische Zahlen derart sind, daß $\log\alpha_1, \ldots, \log\alpha_n$ über den rationalen Zahlen linear unabhängig sind (d. h. kein echtes Potenzprodukt der $\alpha_1, \ldots, \alpha_n$ ist 1!), dann sind $1, \log\alpha_1, \ldots, \log\alpha_n$ sogar linear unabhängig über dem Körper der algebraischen Zahlen. In anderen Worten: Die Linearform $\Lambda = \beta_0 + \beta_1 \log\alpha_1 + \cdots + \beta_n \log\alpha_n$, $\beta_0 \neq 0$, $\beta_0, \ldots, \beta_n$ algebraisch ist immer $\neq 0$. Insbesondere ist $\beta_1 \log\alpha_1 + \cdots + \beta_n \log\alpha_n$ fast immer eine transzendente Zahl. In Kap. 3 wird eine untere Schranke für $|\Lambda|$ hergeleitet. Man betrachtet algebraische Zahlen $\alpha_1, \ldots, \alpha_n, \beta_0, \beta_1, \ldots, \beta_n$ mit beschränktem Grad und kann dann behaupten: Entweder ist $\Lambda = 0$ (diese Fälle übersieht man), oder es ist $|\Lambda| > B^{-C}$, wobei C effektiv aus den $\alpha_1, \ldots, \alpha_n$ und der Grad-Schranke berechnet werden kann. B ist eine obere Schranke für die „Höhe“ der Zahlen $\beta_0, \ldots, \beta_n$. (Die Höhe einer algebraischen Zahl α ist das Maximum der $|a_i|$ in der definierenden Gleichung $a_0\gamma^m + a_1\gamma^{m-1} + \cdots + a_m = 0$ mit *teilerfremden* ganzrationalen $a_0, a_1, \ldots, a_m$.) Die Abschätzungen der Linearformen in Logarithmen liefern bisher die einzigen Methoden, alle ganzzahligen Lösungen (x, y) von Gleichungen der Form $y^2 = f(x) =$ Polynom mit ganzzahligen Koeffizienten (Grad von $f \geqq 3$) (im Prinzip) effektiv zu berechnen (s. Kap. 4). Die Ausdehnung dieser Methoden auf kompliziertere Gleichungstypen stellt ein schwieriges aktuelles Problem dar. — Es ist nicht ausgeschlossen, daß die aus der Theorie der transzendenten Zahlen fließenden Resultate eine Bedeutung erlangen werden, die wesentlich über die Zahlentheorie und die Theorie der diophantischen Gleichungen hinausgehen wird.

Berlin O. NEUMANN

Ameling, W., Laplace-Transformation, Studienbücher Naturwissenschaft und Technik, Bd. 7, Bertelsmann Universitätsverlag. Düsseldorf. 1975. 291 S., DM 29,—.

Der Verf. wendet sich mit diesem Buch vor allem an Studierende der Hoch- und Fachhochschulen sowie an in der Praxis tätige Ingenieure der Elektrotechnik, Elektronik und Regelungstechnik. Der Stoff wird in einer solchen Art und in einem solchen Umfang dargeboten, daß es dem genannten Leserkreis leichtfallen dürfte, sich ein relativ vollständiges Wissen über diese spezielle Integraltransformation anzueignen. An Voraussetzungen werden lediglich Kenntnisse aus Einführungsvorlesungen über Differential- und Integralrechnung sowie über die Grundlagen der Elektrotechnik verlangt. Die nach jedem Kapitel eingefügten, ausführlich durchgerechneten Beispiele vorwiegend aus der Elektrotechnik und Regelungstechnik beweisen die langjährige Erfahrung des Verf. als Hochschullehrer und werden sicherlich mit Gewinn vom Studierenden gelesen.

In den ersten beiden Kapiteln werden der Begriff der Integraltransformation erläutert und durch schrittweisen Übergang von FOURIER-Reihe über FOURIER-Integral zum LAPLACE-Integral das physikalische Verständnis dieser Transformation geweckt. Nach Berechnung einiger einfacher Bildfunktionen zu Beginn des dritten Kapitels findet der Leser eine Reihe von Hilfssätzen zur Überführung von Rechenoperationen des Originalbereichs in einfache arithmetische Operationen des Bildbereichs (Ableitungssatz, Integralsatz, Ähnlichkeitssatz, Dämpfungssatz, Verschiebungssatz, Faltungssatz). Es schließen sich in großer Ausführlichkeit die Methoden der Umkehrung der LAPLACE-Transformation an. Neben dem Gebrauch von Tabellen und der Methode der Partialbruchzerlegung, den beiden wichtigsten Methoden der Rücktransformation, wird auch kurz auf das komplexe Umkehrintegral eingegangen. Bei der Behandlung von Anwendungen der LAPLACE-Transformation wurden bei der Auswahl des Stoffes entsprechend ihrem Einsatz Fragen der elektrischen Netzwerke und Regelungstechnik, des dynamischen Verhaltens und der Simulation technischer Vorgänge in den Vordergrund gestellt. Einen kurzen Einblick in die Anwendung der LAPLACEtransformation zur Lösung partieller Differentialgleichungen am Beispiel der Telegraphengleichung und zur Behandlung von Differenzengleichungen enthalten die letzten Kapitel. Schließlich geht der Verf. noch kurz auf den Zusammenhang zur Operatorenrechnung und zur LAPLACE-CARSON-Transformation ein. Den Abschluß des Buches bildet ein ausgedehnter Tabellenanhang, in dem die wichtigsten Formeln der LAPLACE-Transformation zusammengefaßt, ferner 212 verschiedene Korrespondenzen stetiger und stückweise stetiger Funktionen sowie 49 Kurzschlußimpedanzen und 17 Übertragungs- und Übergangsfunktionen von Verstärkerschaltungen enthalten sind.

Die vorliegende Einführung in das Gebiet der LAPLACE-Transformation unterscheidet sich durch die starke Einbeziehung von Problemen der Elektrotechnik und Regelungstechnik von ähnlichen Buchdarstellungen. Sie kann daher vor allem Studierenden der genannten Fachrichtungen empfohlen werden.

Berlin B. HEIMANN

Ehrig, H., Universal Theory of Automata, Teubner Studienbücher Informatik, 240 S. mit Abb., Stuttgart. B. G. Teubner. 1974. DM 22,80.

In dem vorliegenden Buch werden automatentheoretische Probleme mit Hilfe der Kategorien gelöst, wobei der Schwerpunkt auf einer einheitlichen und allgemeinen Behandlung der verschiedenen Automatentypen liegt. Diese einheitliche Darstellung wird sehr ausführlich im ersten Kap. an Hand der klassischen Automaten motiviert. Anschließend sind im zweiten Kap. die Probleme zusammengestellt, deren Lösung in den folgenden Kap. angestrebt wird. Es handelt sich dabei um die Reduktion, die Minimisierung und die Realisierung eines bestimmten Verhaltens. Wie im ersten Kap. werden diese Probleme ausführlich für die deterministischen und nicht-deterministischen Automaten erläutert. Da sich bekanntlich die Verfahren zur Lösung der o. g. Probleme für beide Automatentypen stark unterscheiden, bedingt dies auch ein differenziertes Vorgehen mit Hilfe kategorientheoretischer Methoden. Deshalb wird in den Kap. 3 bis 5 die „deterministische" Theorie der Reduktion, Minimisierung und Realisierung behandelt. Die sog. abgeschlossenen monoidalen Kategorien bilden den Rahmen für das Studium „deterministischer" Automatentypen, wie z. B. gewöhnlicher deterministischer, partieller, linearer, bilinearer, topologischer u. a. Automaten. Für eine einheitliche und allgemeine Behandlung „nicht-deterministischer" Automatentypen, wie z. B. gewöhnlicher nicht-deterministischer stochastischer, relationaler u. a. Automaten, wird erstmals in diesem Buch der Begriff der sogenannten pseudoabgeschlossenen monoidalen Kategorie verwendet. Damit gelingt es, die o. g. Probleme auch für diesen Fall befriedigend zu lösen (Kap. 6 bis 8). Speziell mit initialen Automaten beschäftigen sich Kap. 9 und 10. Für initiale „nicht-deterministische" Automaten ist ein Minimisierungsverfahren angegeben, das aber nur eine Approximation in bezug auf die Zustandsminimisierung darstellt. Kap. 11 gibt einen Überblick über die Strukturtheorie der Automaten, und im Kap. 12 sind als Anhang die Grundbegriffe der Kategorientheorie zusammengestellt.

Für einen Leser, der an der Anwendung der Kategorientheorie in der Automatentheorie interessiert ist, ist das Buch sehr zu empfehlen.

Dresden W. WECHLER

Pears, A. R., Dimension theory of general spaces, 428 S., Cambridge-London-New York-Melbourne. Cambridge University Press. 1975. £ 16.50 A.

Die Dimensionstheorie ist innerhalb der Topologie ein bis in die jüngste Zeit hinein vielbearbeitetes Gebiet geblieben. Dies mag seinen Grund auch darin haben, daß es *die eine* befriedigende Definition der Dimension eines beliebigen topologischen Raumes nicht gibt, daß aber eben deshalb zuweilen sich recht pathologische und überraschende Resultate einstellen.

Eine Dimensionsfunktion d ist eine Abbildung der Klasse der topologischen Räume in die Menge $G \cup \{+\infty\}$, G Menge der ganzen Zahlen, mit den Eigenschaften

1) $\mathrm{d}(X) = \mathrm{d}(Y)$, sofern die Räume X und Y homöomorph sind,
2) $\mathrm{d}(R^n) = n$ für jedes $n \in N$, wobei R^n der n-dimensionale EUKLIDische Raum sei.

Die wichtigsten Dimensionsfunktionen sind

a) die Überdeckungsdimensionsfunktion dim, die auf eine Erkenntnis von LEBESGUE (1911) zurückgeht, der feststellte, daß jede Überdeckung des Würfels I^n durch abgeschlossene Mengen mit genügend kleinem Durchmesser eine Ordnung $\geqq n$ hat. ČECH (1933) fand die passende Verallgemeinerung für beliebige topologische Räume.
b) die kleine induktive Dimensionsfunktion ind, die auf einer Idee von POINCARÉ (1912) fußt und für die BROUWER (1913) eine exakte Definition gab. Die Theorie von ind wurde von URYSOHN (1922—1926) und von MENGER (1923, 1924) für kompakte metrische Räume entwickelt und von HUREWICZ (1926) und TUMARKIN (1926, 1928) auf die Klasse der separablen metrischen Räume ausgedehnt.
c) die große induktive Dimensionsfunktion Dim, die von ČECH (1932) eingeführt wurde.

Einen bedeutenden Anteil an dem Ausbau der Dimensionstheorie hat u. a. die Schule um P. S. ALEXANDROV. Leider scheint dem Verfasser das schöne Buch П. С. Александров, Б. А. Пасынков „Введение в теорию расмерности (1973) nicht bekannt zu sein, in welchem durch ein geometrisch-anschauliches Beispiel dem lernenden Leser der nicht leicht verständliche Begriff dim plausibel gemacht wird. Für separable metrische Räume fallen die drei genannten Dimensionsfunktionen zusammen. P. ROY konstruierte (1962, 1968) einen metrischen Raum X, so daß ind $X = 0$ und dim $X =$ Ind $X = 1$ ist. Dieses Beispiel wird neben einer Reihe anderer verblüffender — aber bekannter — Beispiele mit dargestellt.

Für kompakte (HAUSDORFF-) Räume X (Verfasser verwendet dafür die etwas veraltete Benennung „bikompakte Räume") gilt allgemein dim $X \leqq$ ind $X \leqq$ Ind X.

Die 1935 von P. S. ALEXANDROV gestellte Frage, ob für kompakte Räume X stets dim $X =$ ind $X =$ Ind X gilt, wurde von LUNC (1949) und LOKUSIEVSKI (1949) negativ entschieden durch Angabe von kompakten Räumen X mit dim $X <$ ind X resp. dim $X <$ Ind X. Offen war bis 1969 die Frage, ob für kompakte Räume stets ind $X =$ Ind X gilt. FILIPPOV fand 1969 einen kompakten Raum X mit der Eigenschaft ind $X <$ Ind X. Dieses Beispiel und auch die Beispielskala von VOPENKA (1948), welche die Beziehungen von dim zu ind und Ind voll aufklärt, kommen in dem Buch zur Darstellung. Ferner werden u. a. für die einzelnen Dimensionsfunktionen sogenannte Teilmengentheoreme

3) $\mathrm{d}(Y) \leqq \mathrm{d}(X)$, falls Y aus einer gewissen Klasse von Teilräumen des topologischen Raumes X ist,

bewiesen. Keinesfalls gilt ein solches Teilmengentheorem für alle Dimensionsfunktionen in voller Allgemeinheit. Dies ist allerdings der Fall für ind, während für Ind und dim i. a. die Abgeschlossenheit von Y zu fordern ist. In dem Buch von PEARS werden hierzu natürlich noch detaillierte moderne Ergebnisse zur Darstellung gebracht. Dies gilt auch für die sogenannten Summentheoreme

4) $\mathrm{d}(X) = \sup_{\alpha \in A} d(Y_\alpha)$, falls $(Y^\alpha)_{\alpha \in A}$ aus einer gewissen Klasse von Überdeckungen von X ist.

Ferner werden behandelt Produkttheoreme, Beziehungen der Dimension von X zur Dimension der STONE-ČECH-Erweiterung βX etc. Es werden ferner noch andere Dimensionsfunktionen behandelt, wie die lokale Dimensionsfunktion locdim und — was sehr erfreulich ist — die KATETOVsche algebraische Charakterisierung (1950) von dim, die über den Ring $C(X)$ der stetigen Funktionen auf X gewonnen wird. Für weitere Einzelheiten muß auf das in sich konsequent und modern aufgebaute Buch verwiesen werden. Das Buch ist wegen eines sehr gut gelungenen rein topologischen Einleitungsabschnittes, der 110 Seiten umfaßt, und u. a. die Metrisationssätze von NAGATA, SMIRNOV und BING enthält, für sich lesbar. Allerdings bedauert der Referent das Abweichen des Verfassers von den längst zu Standardbegriffen gewordenen Begriffen wie normaler Raum etc. Ein normaler Raum ist bei PEARS ein topologischer Raum, in welchem je zwei abgeschlossene disjunkte Mengen durch offene Mengen getrennt werden können. Also braucht ein normaler Raum im Sinne des Verfassers nicht notwendig ein HAUSDORFF-Raum zu sein. Dieses Vorgehen kann bei dem lernenden Leser zu Mißverständnissen führen und erschwert nach Meinung des Referenten auch dem Fortgeschrittenen etwas die Arbeit mit diesem inhaltsreichen Buch.

Greifswald F. TERPE

Kroschel, K., Statistische Nachrichtentheorie, Teil 1, Signalerkennung und Parameterschätzung, Berlin/Heidelberg. Springer-Verlag. 1973. VIII, 183 S., 60 Abb., DM 22,—. US $ 9.10. Hochschultext.

Das vorliegende Buch stellt den ersten Band einer zweiteiligen Einführung in die statistische Nachrichtentheorie dar. Es ist das erklärte Ziel des Verfassers, die auf diesem Gebiet bestehende Lücke in der deutschsprachigen Fachliteratur zu schließen und dem Leser eine leicht lesbare Darstellung in die Hand zu geben. Inhaltlich wendet sich das Buch an Studenten der höheren Semester der Fachrichtungen Nachrichtentechnik und Regelungstechnik sowie an Ingenieure, die aufgrund ihres Arbeitsgebietes einen einführenden Einblick in die statistische Nachrichtentheorie gewinnen wollen. Vorausgesetzt werden Kenntnisse der System- und Netzwerktheorie sowie der Wahrscheinlichkeitsrechnung, deren wesentliche Ergebnisse im 1. Kapitel zusammengefaßt werden. Im 2. Kapitel wird eine auf die Probleme der statistischen Nachrichtentheorie zugeschnittene Darstellungsweise von Signalen mit Hilfe von orthonormierten Basisfunktionen hergeleitet (Vektorkanäle). Diese Darstellung hängt eng mit den kanonischen Zerlegungen von Zufallsprozessen zusammen, ohne daß dieser Zusammenhang erwähnt wird. Die einfache und multiple Detektion mit ihren verschiedenen Optimalitätskriterien (Bayes-Kriterium, Maximum-A-Posteriori-Kriterium, Neymann-Pearson-Kriterium) werden im 3. Kapitel behandelt. Einige Ausführungen zur Realisierung von Korrelationsempfängern mit Hilfe von Matched-Filtern beschließen diesen Abschnitt. Im letzten Kapitel werden die Prinzipien der Parameterschätzung hergeleitet. Besonders Interesse finden dabei die optimalen linearen Schätzeinrichtungen, deren Beschreibung durch das Gauß-Markow-Theorem erfolgt. Das für die lineare Schätztheorie fundamentale Orthogonalitätsprinzip wird in diesem Zusammenhang gut herausgearbeitet. Den Abschluß dieses Kapitels bilden die bekannten Kalman-Formeln zur Verbesserung der Schätzung bei Kenntnis zusätzlicher Meßdaten.

Die bei der Darstellung des gebotenen Stoffes vielfach verwendeten Tabellen und Zusammenfassungen unterstreichen den einführenden Charakter des Werkes. Sie sind ebenso eine Hilfe für den Studierenden wie die Aufgabensammlung des Buches, in der zu jedem Kapitel typische Aufgaben einschließlich ihrer Lösungen enthalten sind. Einige kleine Unkorrektheiten bei der Abfassung des Textes, namentlich bei der Erklärung von Formelgrößen sollten überarbeitet werden, so bleiben beispielsweise die Begriffe der bedingten Wahscheinlichkeit (S. 13) und die der Diracfunction (S. 16) unerklärt. Auf S. 26 verwirren unklare Indizes das Verständnis der dort abgeleiteten Integralgleichung zur Ermittlung der Basisfunktionen.

Berlin B. HEIMANN

Kroschel, K., Statistische Nachrichtentheorie, Teil 2, Signalschätzung, Berlin/Heidelberg. Springer-Verlag. 1974. VII, 189 S., 41 Abb., DM 23,—. US $ 9.40.

Im zweiten Band der zweiteiligen Einführung in die statistische Nachrichtentheorie werden zusammenfassend alle Aufgaben der linearen Signalschätzung behandelt, d. h. Filterung, Prediktion und Interpolation zweitdiskreter und und kontinuierlicher stationärer und instationärer Prozesse. Der Verf. knüpft dabei an die Ergebnisse vor allem der Parameterschätzung vom ersten Teil an. Da diese Ergebnisse jedoch kurz wiederholt werden, ist das vorliegende Buch auch ohne Kenntnis des ersten Teils lesbar. Vorausgesetzt werden, wie beim ersten Teil, Kenntnisse der System- und Netzwerktheorie sowie der Statistik.

Die Betrachtungen werden auf lineare Systeme mit quadratischer Gewichtung des Schätzfehlers beschränkt, weshalb ausschließlich der mittlere quadratische Fehler als Optimalitätskriterium herangezogen wird. Das daraus resultierende Orthogonalitätsprinzip wird als entscheidende Beziehung für die Signalschätzung verwendet; aus ihm werden alle im Buch behandelten optimalen Systeme hergeleitet.

Im 1. Kapitel werden die Schätzung zeitkontinuierlicher und zeitdiskreter stationärer Prozesse mit Hilfe der Wiener-Filter behandelt, die dazugehörende Wiener-Hopf-Integralgleichung abgeleitet und auf ihre Lösungsmöglichkeiten eingegangen. Als Grundlage für die Schätzung instationärer Prozesse wird im 2. Kapitel ein universelles Modell abgeleitet, das auf der Beschreibung von Systemen durch Zustandsvariable aufbaut. Die zur Signalschätzung instationärer wie stationärer zeitdiskreter Prozesse bekannten Kalman-Filter sind Gegenstand des 3. Kapitels. Als Kernlösung dieses Abschnitts werden für die Prediktion um einen Schritt rekursive Schätzformeln hergeleitet. Ausgehend von diesen Ergebnissen werden Schätzformeln für die allgemeine Prediktion, für die Filterung und die Interpolation angegeben und schließlich der enge Zusammenhang dieser Ergebnisse mit denen der sequenziellen Parameterschätzung herausgearbeitet.

Im letzten Kapitel wird die Schätzung zeitkontinuierlicher instationärer Prozesse mit Hilfe der Kalman-Bucy-Filter behandelt.

Zur besseren Untergliederung und Zusammenfassung des Stoffes dienen wieder zahlreiche Tabellen, in denen nicht nur die wichtigsten Ergebnisse zusammengestellt, sondern auch Zwischenrechnungen, die den Ablauf der Herleitung unterbrechen würden, vom Text getrennt dargestellt sind. Den Schluß des Buches bildet wieder ein Aufgabenteil, in dem typische Aufgaben und ihre Lösungen enthalten sind. Die Herleitung der Schätzformeln erfolgt in der Regel durch elementare Rechenoperationen und im Hinblick auf apparative Realisierungen. Um den einführenden Charakter des Werkes zu unterstreichen, wurden bewußt Wiederholungen bei der mathematischen Behandlung einzelner Anwendungsfälle, z. B. der Filterung und der Prediktion, vorgenommen.

Das Buch kann vor allem Studenten und Ingenieuren der Fachrichtungen Nachrichtentechnik und Regelungstechnik empfohlen werden, die sich mit den Methoden und den Ergebnissen der linearen Signalschätzung vertraut machen wollen.

Berlin B. HEIMANN

EINGEGANGENE BÜCHER

Bei der Schriftleitung sind folgende Bücher eingegangen (Besprechung bleibt vorbehalten).

Berkovitz, L. D., Optimal Control Theory. Berlin-Heidelberg-New York. Springer-Verlag. 1974. 304 S., 10 Abb., DM 23,30. (Appl. Mathematical Sciences 12)

Taylor, M., Pseudo Differential Operators. Berlin-Heidelberg-New York. Springer-Verlag. 1974. 155 S., DM 18,–. US $ 7.40. (Lecture Notes in Mathematics 416).

Boyce, D. E. / Farhi, A. / Weischedel, R., Optimal Subset Selection. Multiple Regression, Interdependence and Optimal Network Algorithms. Berlin-Heidelberg-New York. Springer-Verlag. 1974. 187 S., DM 20,–. US $ 8.70. (Lecture Notes in Economics and Mathem. Systems 103).

Ludwig, D., Stochastic Population Theories. Notes by Levandowsky, M., Berlin-Heidelberg-New York. Springer-Verlag. 1974. 108 S., DM 18,–. US $ 7.90. (Lecture Notes in Biomathematics 3)

Saletan, E. J. / Cromer, A. H., Theoretische Mechanik. München-Wien. R. Oldenbourg Verlag. 1974. 395 S., 41 Abb., M 64,–.

Feynman, R. P./Leighton, R. B./Sands, M., Vorlesungen über Physik. Band I: Mechanik Strahlung Wärme, Tl. 1, Tl. 2, Band II: Elektromagnetismus und Materie, Tl. 1, Tl. 2, Band III: Quantenmechanik, München-Wien. R. Oldenbourg Verlag. 1971.

Kossack, C. F./Henschke, C. I., Introduction to Statistics and Computer Programming. San Francisco. Holden-Day, Inc., 1975. 651 S., $ 10.95.

Herschel, R., Einführung in die Theorie der Automaten, Sprachen und Algorithmen. München-Wien. R. Oldenbourg Verlag. 1974. 226 S., 152 Abb., DM 29,–.

Roseau, M., Asymptotic Wave Theory. Amsterdam-Oxford. North-Holland Publ. Company. 1976. X u. 349 S., Dfl. 75.00 US $ 29.50. (Appl. Mathematics and Mechan. Bd. 20).

Sinclair, A. M., Automatic Continuity of Linear Operators. Cambridge-London-New York-Melbourne. Cambridge University Press. 1976. 93 S., £ 2.80 A. (London Mathematical Society Lecture Notes Series 21).

NACHRICHTEN

Ludwig Föppl †

Ludwig Föppl, der am 13. 5. 1976 im neunzigsten Lebensjahr verstorben ist, wirkte von 1925 ab über drei Jahrzehnte als Professor für Technische Mechanik an der Technischen Universität München. Dank seiner starken persönlichen Ausstrahlung und seiner großen pädagogischen Begabung verstand er es, den Studierenden die Erkenntnisse dieses grundlegenden Fachgebiets überzeugend und eindrucksvoll zu vermitteln.

Nachdem sein berühmter Vater, August Föppl, nach München berufen worden war, studierte Ludwig Föppl an der Technischen Hochschule und an der Universität München. Später promovierte er an der Universität Göttingen und habilitierte sich an der Universität Würzburg. Nach dem ersten Weltkrieg übertrug ihm die TH München bereits einige Vorlesungen. 1920 wurde er als Ordinarius für Mechanik an die TH Dresden berufen. Als sein Vater 1922 emeritiert wurde, berief man ihn wieder an die TH München, wo er bis zu seiner Emeritierung 1956 glänzende Vorlesungen hielt. Ludwig Föppl war dreimal Dekan. In der besonders schwierigen Zeit nach dem zweiten Weltkrieg leitete er als mehrjähriger Rektor den Wiederaufbau der TH mit großem Geschick. In Anerkennung seiner Verdienste wurden ihm zahlreiche Ehrungen und Auszeichnungen zuteil. So war er ordentliches Mitglied in der Bayerischen Akademie der Wissenschaften, korrespondierendes Mitglied in der Spanischen Akademie der Wissenschaften und im Mathematischen Zirkel zu Palermo sowie Träger des Bundesverdienstkreuzes und des Bayerischen Verdienstordens. Zu seinen Schülern zählen bedeutende Ingenieure in Industrie und Forschung.

Das sechsbändige Standardwerk „Vorlesungen über Technische Mechanik" seines Vaters wurde von Ludwig Föppl neu bearbeitet. Es erreichte in Deutschland 15 Auflagen und wurde in viele Fremdsprachen übersetzt. Zusammen mit seinem Vater schrieb er ein zweibändiges Werk über die Elastizitätstheorie, das er später um einen Band über den ebenen Spannungszustand erweiterte. Zahlreiche wissenschaftliche Veröffentlichungen zeigen, daß Ludwig Föppl bei der rechnerischen Behandlung technischer Probleme aus der Mechanik — besonders der Festigkeitslehre und der Kinematik — bahnbrechend gewirkt hat. Sein Ziel war stets, möglichst einfache Lösungen zu finden, die dem Ingenieur die wesentlichen Zusammenhänge deutlich machten.

Unter seiner Leitung sind aus dem von Bauschinger gegründeten Mechanisch-Technischen Laboratorium in München bedeutende, für die Materialprüfung und Festigkeitsforschung richtungweisende Arbeiten hervorgegangen. Das gilt ganz besonders für die Spannungsoptik, die er wesentlich weiterentwickelt hat. Gemeinsam mit H. Neuber schrieb er das Buch „Festigkeitslehre mittels Spannungsoptik", mit E. Mönch „Praktische Spannungsoptik" und mit G. Sonntag „Tafeln und Tabellen zur Festigkeitslehre". Nach seiner Emeritierung widmete er sich weiterhin der Spannungsoptik, dann aber auch der historischen Entwicklung der Grundgedanken der Mechanik. Viel Beachtung fand sein Buch „Elementare Mechanik vom höheren Standpunkt", in dem er seine Gedanken zur erkenntnistheoretischen Entwicklung der Grundlagen der Mechanik vom Altertum bis zur Neuzeit darlegte.

H. Neuber

Heinz Neuber — 70 Jahre

Wer je mit Methoden oder Ergebnissen der modernen Elastizitätstheorie und Festigkeitslehre in Berührung gekommen ist, kennt die Arbeiten von H. Neuber. Sie gehören zweifellos zu den am häufigsten zitierten Arbeiten des engeren Fachgebietes, da sie eine sichere Grundlage für die Lösung spezieller Festigkeitsprobleme bilden und Ansatzpunkte für eine Weiterentwicklung der hier eingesetzten theoretischen Methoden bieten.

Der seit kurzem emeritierte ordentliche Professor für Mechanik, Dr.-Ing., Dr. rer. nat. h. c. Heinz Neuber, vollendet am 22. November 1976 sein 70. Lebensjahr. Im Jahre 1906 in Stettin geboren, studierte Neuber in den Jahren 1925—1929 an den Technischen Hochschulen in Berlin und München. Nach dem Abschluß des Studiums arbeitete er bis 1935 in München als Assistent und enger Mitarbeiter von Ludwig Föppl. In diese Zeit fallen sowohl seine Promotion über „Beiträge für den axialsymmetrischen Spannungszustand" (1932) als auch die Habilitation für das Fach „Mechanik" mit einer Arbeit „Der räumliche Spannungszustand in Umdrehungskerben".

Nach zweijähriger Tätigkeit in der Industrie (1935—1937) übernahm Neuber die Leitung einer Abteilung bei der Luftfahrt-Forschungsanstalt Völkenrode. Im Jahre 1946 folgte er einem Ruf auf den Lehrstuhl für Technische Mechanik an der Technischen Hochschule Dresden. Hier entfaltete er eine überaus fruchtbare Lehrtätigkeit, die ihren Niederschlag u. a. in der Herausgabe eines von Studenten und Dozenten viel verwendeten Übungsbuches mit Lösungen zu der umfangreichen Aufgabensammlung von Mestscherski fand. Im Jahre 1955 wurde ihm der durch die Eremitierung von Ludwig Föppl freigewordene Lehrstuhl für Mechanik an der Technischen Hochschule München angeboten. Er nahm diesen ehrenvollen Ruf an und hatte damit die schwierige Aufgabe übernommen, in den Fußstapfen so erfolgreicher Vorgänger wie Bauschinger, August und Ludwig Föppl die Arbeiten des traditionsreichen und weltweit bekannt gewordenen Lehrstuhls weiterzuführen. Auch das bereits 1870 von Bauschinger begründete Mechanisch-Technische Laboratorium sowie das Staatliche Material-Prüfungsamt für den Maschinenbau gehörten zu diesem Lehrstuhl.

Insgesamt 6 Fachbücher und über 80 wissenschaftliche Veröffentlichungen zeugen von der erfolgreichen Forschungstätigkeit Neubers. Bereits 1935 gab er gemeinsam mit seinem Lehrer L. Föppl das Werk „Festigkeitslehre mittels Spannungsoptik" heraus. Ein weltweites Echo brachte ihm die 1937 erschienene „Kerbspannungslehre" ein, die ins Englische, ins Russische und ins Japanische übersetzt wurde. Dieses Buch war viele Jahre führend in seiner Art, wie auch aus der Tatsache deutlich wird, daß 20 Jahre nach der Erstveröffentlichung eine Neuauflage notwendig wurde. Der außergewöhnliche Wert des Buches kann wohl darin gesehen werden, daß hier einerseits dem Konstrukteur ein umfassendes, theoretisch fundiertes Rüstzeug zur Berechnung komplizierter Bauteile geboten wurde, andererseits aber auch die experimentelle und theoretische Festigkeitsforschung bis hin zur Bruchmechanik Anregungen erhielt. In Fortführung dieser Arbeiten ist die theoretische Behandlung von Sonderproblemen der Spannungskonzentration bei linearem oder nichtlinearem Stoffgesetz auch weiterhin Gegenstand der Forschungsarbeiten Neubers gewesen. Dabei konnte gerade in den letzten Jahren ein besonderer Fortschritt durch die Theorien der Makro- und Mikro-Stützwirkung erzielt werden.

Neubers Veröffentlichungen beschränken sich freilich nicht nur auf Probleme der Spannungskonzentration. Es gibt kaum ein Gebiet der modernen Festigkeitslehre, zu dem er nicht auch Beiträge geliefert hätte. So gibt es Veröffentlichungen zur Anwendung numerischer Methoden in der Festigkeitslehre, Erweiterungen des Drei-Funktionen-Ansatzes, Arbeiten zu Problemen der elastischen Stabilität, der Plasto-Mechanik, der Thermo-Elastizität, Anwendungen der Theorie des Cosserat-Kontinuums, Untersuchungen über Schwingungen linear elastischer Systeme sowie eine Reihe von Spezialarbeiten zur Theorie der Zahnräder.

Die umfangreiche Lehrtätigkeit Neubers fand ihren Niederschlag in dem dreibändigen Lehrbuch „Technische Mechanik". Dieses primär aus den üblichen Grundvorlesungen hervorgegangene Werk geht freilich nach Inhalt und Darstellungsart erheblich über das Niveau der Anfängervorlesungen hinaus und kann auch für Fachleute als eine wertvolle Fundgrube betrachtet werden. Der Grundgedanke des Lehrbuches war es, eine systematische Darstellung zu geben, die einen zur Mitarbeit bereiten Leser anregt und dadurch geeignet ist, sein technisches Denkvermögen zu schulen. Das ist zweifellos erreicht worden, obwohl — oder vielleicht gerade weil — die Darstellung an entscheidenden Stellen vom Konventionellen so abweicht, daß ein teilweise neuartiger Zugang zu den Grundlagen der Mechanik aufgezeigt wurde.

Neubers weitgespanntes Wirken fand vielfältige Anerkennung, die in zahlreichen Einladungen zu Vorträgen auf internationalen Kongressen, aber auch in der Verleihung der Würde eines Ehrendoktors durch die Bergakademie Freiberg (1961) zum Ausdruck kam. Zwei Akademien, die Berliner Akademie der Wissenschaften und die Bayerische Akademie der Wissenschaften ernannten Neuber zu ihrem Mitglied. Sein Wirken, in eigenen Veröffentlichungen eindrucksvoll dokumentiert, strahlt aber auch über Aktivitäten und Erfolge einer beachtlichen Schar von Schülern und Mitarbeitern aus. Die Fachwelt ist Neuber für das Geleistete Dank schuldig und wünscht ihm zugleich von Herzen Kraft und Muße für ein weiteres erfolgreiches Schaffen.

K. Magnus

FSC
www.fsc.org
MIX
Papier | Fördert
gute Waldnutzung
FSC® C083411

Zeitfracht Medien GmbH
Ferdinand-Jühlke-Straße 7
99095 Erfurt, Deutschland
produktsicherheit@kolibri360.de